Formal Methods in Computer Science

Textbooks in Mathematics

Series Editors
Al Boggess and Ken Rosen

Formal Methods in Computer Science

Jiacun Wang

William Tepfenhart

CRC Press

Taylor & Francis Group

Boca Raton London New York

CRC Press is an imprint of the
Taylor & Francis Group, an **Informa** business

A CHAPMAN & HALL BOOK

CRC Press
Taylor & Francis Group
6000 Broken Sound Parkway NW, Suite 300
Boca Raton, FL 33487-2742

© 2020 by Taylor & Francis Group, LLC
CRC Press is an imprint of Taylor & Francis Group, an Informa business

No claim to original U.S. Government works

Printed on acid-free paper

International Standard Book Number-13: 978-0-367-22570-4 (Hardback)
978-1-4987-7532-8 (Paperback)

Library of Congress Cataloging-in-Publication Data

Names: Wang, Jiacun, 1963- author. | Tepfenhart, William M., author.
Title: Formal methods in computer science / Jiacun Wang and William
Tepfenhart.
Description: Boca Raton : Taylor & Francis, a CRC title, part of the Taylor &
Francis imprint, a member of the Taylor & Francis Group, the academic
division of T&F Informa, plc, 2019.
Identifiers: LCCN 2019004989 | ISBN 9781498775328 (paperback : acid-free
paper)
Subjects: LCSH: Formal methods (Computer science)
Classification: LCC QA76.9.F67 W36 2019 | DDC 004.01/51--dc23
LC record available at https://lccn.loc.gov/2019004989

Visit the Taylor & Francis Web site at
http://www.taylorandfrancis.com

and the CRC Press Web site at
http://www.crcpress.com

Contents

Preface

Formal methods are techniques for the specification, development, and verification of software and hardware systems. A formal method is a mathematical method. By building a mathematically rigorous model of a system, it is possible to verify the system's properties in a more thorough fashion than empirical testing. It is believed that applying formal methods in system design and development will increase the reliability and robustness of the system.

Formal methods can be applied at various points throughout the system development process. The entry step is formal specification, a process that translates nonmathematical description (diagrams, tables, or natural language statements) into a formal specification language, such as Petri nets and temporal logic. A formal specification represents a concise description of a system's high-level behavior and properties. The development of a formal specification provides insights and an understanding of the system requirements and design. It helps to clarify customers' requirements and reveals and removes ambiguities, inconsistency, and incompleteness. It also facilitates unambiguous communication of requirements among stakeholders. A formal specification, once produced, may be used as a guide while the concrete system is developed during the design process. The specification may also be used as the basis for proving or verifying properties of the system under development. Proofs can be done manually, step-by-step, using established proof or deduction rules. Proofs can also be constructed with automated means. Model checking is one of such means. It is a technique that relies on building a finite model of a system and checks whether a desired property holds in that model.

Use of formal methods, however, requires a sound mathematical knowledge of the developer. Oftentimes, one single formal language is not enough for the entire system design, development, and verification process. Different aspects of a design may be represented by different formal specification methods.

Book Layout

This textbook contains 10 chapters. Chapters 1 through 3, and 8, were written by Dr. William Tepfenhart, and Chapters 4 through 7, 9, and 10, were written by Dr. Jiacun Wang. The 10 chapters are organized in three parts.

The first part introduces some fundamentals in formal methods. It has three chapters. Basic set definitions, set operations, ordered pairs, relations, and functions are introduced in Chapter 1. Finite state machines, including Mealy machines, Moore machines, and Harel machines (UML) are introduced in Chapter 2. Strings and languages, regular expressions, Lex, and grammar are presented in Chapter 3.

The second part focuses on logic, a powerful formal language in specifying system properties. It takes into account syntactically well-formed statements and studies whether they are semantically correct. This part is composed of four chapters. Propositional logic, which deals with declarative sentences, or propositions, and is concerned with the study of the truth value of propositions and how their value depends on the truth or falsity of their component propositions, is covered in Chapter 4. Predicate logic, also called first-order logic or first-order calculus, assumes the world contains objects. Objects are represented with variables. The relations between objects are specified with predicates. Predicates, quantifiers, syntax, semantics, and deduction rules in predicate logic are introduced in Chapter 5. Temporal logic is a formal system for specifying and reasoning about a system's dynamic properties. Linear temporal logic (LTL), computation tree logic (CTL), CTL*, and their syntax, semantics, and property specifications are discussed in Chapter 6. Model checking is an automatic verification technique for finite-state concurrent and reactive systems. It is developed to verify if assertions on a system are true or false. The labeling algorithm for CTL model checking and the NuSMV model checking tool and its associate descriptive language are presented in Chapter 7.

Chapters 8 through 10 constitute the third part of the book. This part presents the most popular formal language in system behavior modeling, Petri nets. The fundamentals of Petri nets, including definitions, transition firing rules, reachability analysis, various structural and behavioral properties, reduction rules, and modeling practice are presented in Chapter 8. Timed Petri nets, which are an augmentation of regular Petri nets with timing parameters that model event durations in a system, are introduced in Chapter 9. Deterministic timed Petri nets and stochastic Petri nets are discussed in detail in this chapter. Colored Petri nets are a kind of high-level Petri net, in which each token is attached a color, indicating the identity of the token (object), each place is attached a set of colors, and each transition is associated with a set of bindings that specify the colors of tokens to take from and put into places when the transition fires. These changes lead to much more compact net models. Colored Petri nets are introduced in Chapter 10.

Audience

The book is primarily written as a textbook for undergraduate- or graduate-level courses in computer engineering, software engineering, computer science, and information technology programs in the subject of formal methods and their application in software and hardware specification and verification.

Portions of the book can be used as reading material for undergraduate computer engineering, computer science, and software engineering capstone courses, or as a reference for students who are conducting research in the area of formal system specification and validation.

The book is also useful to industrial practitioners with software system modeling and analysis responsibilities.

Jiacun Wang, PhD
Monmouth University
New Jersey

Acknowledgments

First, I want to thank my family for their patience and the support they have shown while I was working on the book. I wish to thank Dr. Ken Rosen for his encouragement and support in planning this project. I am appreciative of the great help of Ms. Jiani Zhou, a master's degree student of Monmouth University, in preparing some figures and tables in the book.

I also want to thank everyone at CRC Press/Taylor & Francis who worked so diligently on this book. Their dedication and hard work has helped bring the book to you in a timely manner and with the best presentation quality.

I will never forget Prof. William Tepfenhart, the second author of the book, for his great contribution to the work and also for his friendship for nearly 15 years. He was a very knowledgeable and kind-hearted person and was well respected by his colleagues and students. Unfortunately, he succumbed to health issues that he struggled with for months. He will be forever missed by all those who knew him.

Authors

Jiacun Wang, PhD, received a PhD in computer engineering from Nanjing University of Science and Technology (NJUST), China, in 1991. He is currently a professor of software engineering at Monmouth University, West Long Branch, New Jersey. From 2001 to 2004, he was a member of the scientific staff with Nortel Networks in Richardson, Texas. Prior to joining Nortel, he was a research associate at the School of Computer Science, Florida International University (FIU) in Miami. Prior to joining FIU, he was an associate professor at NJUST.

His research interests include formal methods, software engineering, discrete event systems, and real-time distributed systems. He has authored *Timed Petri Nets: Theory and Application* (Kluwer, 1998), *Handbook on Finite-State Based Models and Applications* (CRC Press, 2012), and *Real-Time Embedded Systems* (Wiley, 2017); and has published more than 80 research papers in peer-reviewed international journals and conferences. He was an associate editor of *IEEE Transactions on Systems, Man and Cybernetics, Part C*; and is currently an associate editor of the *International Journal of Discrete Event Systems*. He has served as general chair, program chair, program co-chair, special sessions chair, or program committee member for many international conferences. Dr. Wang has been teaching formal methods for both undergraduates and graduates at Monmouth University for 15 years.

William Tepfenhart, PhD, received a PhD in physics from the University of Texas at Dallas in May (1956–2019). His experience ranged across a broad spectrum of activities. He performed in the role of instructor, researcher, software developer, and author. Trained as a physicist, his areas of expertise included object-oriented software development, artificial intelligence, and software engineering. His knowledge of modeling physical systems formed the basis for major contributions in the area of software development. He was active in the area of OOA/OOD for several large systems. His expertise in this area has been documented in a book titled, *UML and C++: A Practical Guide to Object Oriented Development*.

Dr. Tepfenhart's industrial responsibilities included: preparation of research proposals, system engineering, program development, project management, architect, and basic research. As an active member of the research community, he served as the program chair for the Second International Conference on Conceptual Structures held in Maryland in 1994. He also served as the program chair for the Sixth International Conference on Conceptual Structures held in Virginia in 1999.

1

Set Theory and Functions

The concepts of a set and that of a function are well known to most people in an informal manner. After all, it doesn't take a mathematical genius to talk about a tea set as having four cups, four saucers, a tea pot, a sugar bowl, and a creamer. Anyone who has studied basic algebra in high school knows that the square function applied to a number x produces a number y such that x *times* $x = y$. These are nice informal everyday understandings for sets and functions. However, this informal understanding is not sufficient to generate unambiguous and detailed specifications for use in describing a program.

To be useful in specifying programs and systems, a deeper and more formal understanding of these two concepts is necessary. It requires the clarity, specificity, and precision of mathematics to unambiguously specify the procedures, operations, and functions of a computer program.

In this chapter, set theory and functions are introduced and defined. The introduction uses simple language to define the key concepts in a naïve formalism, rather than a strictly logical formal definition. The formal definition of sets and functions requires symbolic logic, which is introduced in a later chapter. Sets and functions will be revisited using a logic-based formalism. This chapter addresses two basic questions:

1. What is a set?
2. What is a function?

Examples of where sets and functions can be used in specifying program behavior will be given.

1.1 Basic Set Definitions

The exploration of set theory begins by establishing a simple vocabulary that allows for discussion of the concepts without an excessive burden of mathematics. It slowly develops the mathematics of sets as the chapter progresses.

A *set* is an *abstract collection* of *distinct objects*. A set, denoted by {}, exists as an abstraction for some collection of objects where the collection is treated as a whole. It is an object in its own right, albeit an abstract object. For example, a flock of birds (i.e., a set of birds) treats the entire collection of birds as a single entity although it is made up of individual birds.

An object that is part of a collection is an *element* or *member* of the set. The terms element and member are used interchangeably. This book gives preference to the term member. A member of a set is shown within the curly braces. The set containing the three colors RED, GREEN, YELLOW is {RED, GREEN, YELLOW}. The curly braces indicate the set and the values between the curly braces are the members of that set. The member can be an abstraction (e.g., a word, number, or even another set) or a concrete thing (e.g., car, plane, or teacup).

Membership in a set is an all-or-nothing kind of deal. A thing is either a member of the set or it isn't. It can't be a partial member or be a member more than once. However, a thing can be a member of multiple sets. A flock of birds, since it is a set, obeys the constraint that any specific bird at any particular time only appears within the flock once. Basically, a bird can't be in the flock two times at the same time. It's just not possible.

When it is necessary to refer to a set as an entity, the convention is to denote it using a single capital letter such as A. Also, according to convention, a lowercase letter is used to refer to a member of a set. To identify one individual among a number of set members, set members are often denoted by the first few lowercase letters or a lowercase letter with a subscript to distinguish among the members. For example, one could have {a, b, c} or {a_1, a_2, a_3}.

The establishment of a co-reference between the name and the set is accomplished by using an assignment operator, =, in the manner A = {a_1, a_2, a_3}. When several sets are being discussed simultaneously, they are often denoted by the first few capitals: A, B, C, and so forth. In some cases, a subscript might be employed if the sets are of the same type but containing different members. In referring to an arbitrary, generic set, a typical notational choice is S.

The use of a Venn diagram is a common way to illustrate sets and operations over sets. A Venn diagram has a rectangular area that represents the universal set, U. One can view the universal set as the set of everything that matters in this problem space. A set is drawn as an enclosed area within the rectangle. Members of the set are within the boundary while objects outside the boundary are not.

Example 1.1

A simple Venn diagram is shown in Figure 1.1. This diagram shows a set, labeled A, that contains a, b, and c out of a universe, U, that contains a, b, c, and d.

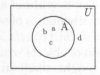

FIGURE 1.1
Example Venn diagram.

The identity of a set is established by the members of the set. The order in which members appear in the written description of a set is not significant. There is not a concept of order within a set.

Example 1.2

Figure 1.1 shows a set containing three members, a, b, and c. All the following forms of establishing the identity of the set are equivalent and identify exactly the same set, A,

$$A = \{a, b, c\}$$
$$A = \{a, c, b\}$$
$$A = \{b, a, c\}$$
$$A = \{b, c, a\}$$
$$A = \{c, a, b\}$$
$$A = \{c, b, a\}$$

(1.1)

Referring back to Figure 1.1, there isn't anything in that figure to suggest a specific order for the members of A.

Rather than having to always list the members of set A in the form, A = {a, b, c}, it is possible to use a membership operator, $\in$. To show that a is an element of A, one would write a $\in$ A. The converse (negation) of the membership operator is $\notin$. To show that b is not an element of B, one would write b $\notin$ B. Identity and membership are important when discussing the operations that can be performed over sets.

Example 1.3

The following expressions illustrate the two operators applied to set A that was shown in Figure 1.1.

$$a \in \{a, b, c\} \qquad a \in A$$
$$b \in \{a, b, c\} \qquad b \in A$$
$$c \in \{a, b, c\} \qquad c \in A$$
$$d \neg \{a, b, c\} \qquad d \neg A$$

(1.2)

The use of membership operators is oftentimes the easiest way of specifying membership to a set.

A set can be specified by identifying its members in two ways: by *extension* or by *intension*. In specifying a set by extension, one identifies the members explicitly. For example, {Mercury, Mars, Earth, Venus, Jupiter, Saturn, Neptune, Uranus} is the set of planets with each member explicitly identified by name. In specifying a set in this manner, it is important to understand that the members are the objects named by the text, not the text itself. One can use a convention that {Pluto} is the set containing the former planet named Pluto, {"Pluto"} is the set containing the name Pluto, and {"Pluto"} is the set containing the string Pluto.

While it might seem reasonable to identify each person in a class, such as a formal methods class, individually, it is not so convenient to identify every person in the world individually. One could spend a lifetime trying to do that. Instead, it makes sense to specify large sets using intension. When specifying a set by intension, one identifies some criteria that determines whether any given object is a member or not. This can be done in the form of a predicate or a generating function.

The notation for identifying members by intension uses braces, {}, to denote the set and with a bar separating a member variable and the condition, which is used to determine if any given thing is a member or not. The notation is

$$\{item \,|\, \text{condition}(item)\} \tag{1.3}$$

This can be read as the set of all *item*(s) such that each *item* satisfies condition. The variable, in this case item, is used as a temporary identifier for an object and the condition is some logical expression that returns true or false depending on whether or not the item satisfies the condition. For now, a simple pseudo-code mechanism for specifying the logical expression will be used. In a later chapter, a formal means for specifying the condition as a predicate using logic will be described in detail.

Example 1.4

Suppose one needs to discuss the set of all yellow cats. It would be very difficult to list each member individually so one would specify it by intension. The set containing all yellow cats would be specified as

$$\{x \,|\, x \text{ is a cat AND } x \text{ is yellow}\} \tag{1.4}$$

This would be read as the set of all x such that x is a cat and x is yellow. Any item that is considered for membership to this set can be accepted or rejected by asking, "Is it a cat?" and "Is it yellow?" If the answer to both of those questions is yes, then it is a member. Otherwise, it is not a member.

An alternative way of specifying membership is to use a recursive rule to identify members. This operates by using some members as a seed by which other members can be computed.

Example 1.5

Specifying a set containing the counting numbers can be accomplished recursively. The algorithm is

$$1 \in S$$

$$\text{if } x \in S, \text{ then } x + 1 \in S \tag{1.5}$$

$$\text{nothing else belongs to S}$$

The first line establishes that 1 is a member of the set. The second line establishes that 2, then 3, then 4, and so on, are members of the set. The last line says that anything else, which isn't generated by the first two rules, is not a member of the set. By this rule, 1.5 is not a member of the set.

One can study sets and how sets behave independent of the kind of objects that are its members. One important concept is that of cardinality. The *cardinality* of a set is the number of members within the set. The cardinality of the set S is denoted by enclosing the set within a pair of bars, $||$. The expression $|S|$ denotes the cardinality of the set S.

Sets can have zero members, one member, a finite number of members, or an infinite number of members. A set with zero members, such as the set containing all living unicorns, is called a *null set*, and is written as $\{\}$, or alternatively, $\emptyset$. A set with a single member, such as the set containing all suns around which the earth orbits, is called a *singleton*. The set of planets in the solar system is a *finite* set. The set of integer numbers is an *infinite* set.

Example 1.6

The Venn diagram of Figure 1.2 shows three sets (four, if one includes the universal set):

$$A = \{a, b, c\}$$

$$B = \{c, d, e, f\} \tag{1.6}$$

$$C = \{\}$$

$$U = \{a, b, c, d, e, f, g\}$$

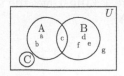

FIGURE 1.2
Example Venn diagram for three sets.

The cardinality of these sets is given by:

$$|A| = 3$$

$$|B| = 4$$

$$|C| = 0 \tag{1.7}$$

$$|U| = 7$$

The universal set U was included because it establishes the things that are important in this problem space, namely a, b, c, d, e, f, and g. Normally, one does not explicitly identify the members of the universal set.

As a shorthand notation, when there is a well-known sequencing among members, such as one would have with the integers 1 through 8, one could represent it is as {1,...,8}, where the ellipsis stands for the members that are not explicitly represented, but fall between the two members in the recognized sequence. A set with infinite members, such as the set of counting numbers could be denoted as {1, 2,...}, while the set of integers could be denoted as {..., -2, -1, 0, 1, 2,...}.

Two important concepts associated with sets is that of subset and superset. Set A is a *subset* of set B if every member of A is also a member of B. This relationship between A and B is denoted using the subset symbol, $\subseteq$, as

$$A \subseteq B \tag{1.8}$$

One reads this as A is a subset of B.

The set B is a *superset* of set A if every member of A also appears in B. This relationship between A and B is denoted using the superset symbol, $\supseteq$, as

$$B \supseteq A \tag{1.9}$$

One reads this as B is a superset of A.

It should be obvious that subsets and supersets are related to each other as follows:

$$\text{If A is a subset of B, then B is a superset of A.} \tag{1.10}$$

The Venn diagram of Figure 1.3 illustrates the subset and superset concept with B being a superset of A.

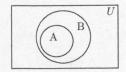

FIGURE 1.3
Subset and superset examples.

Example 1.7

Given the following sets:

$$A = \{a, b, c\} \tag{1.11}$$

$$B = \{a, b, c, d\}$$

one can say that the following statements are true:

$$A \subseteq B \tag{1.12}$$

$$B \supseteq A$$

since every member of A, namely a, b, and c, are also members of B. B is not a subset of A since it contains the member d, which is not in A.

It should be noted that any given set S is a subset of itself since it contains every member within it. Thus, $S \subseteq S$ is always true. It is often important to make a distinction between a subset that is identical to the superset and the case in which the superset has members that aren't in the subset.

A *proper subset* is a set A that only contains elements of B, but not all elements of B. This is denoted using the symbol $\subset$, as

$$A \subset B \tag{1.13}$$

A set S cannot be a proper subset of itself. In the same way that there is a proper subset, there is a proper superset which is denoted using the symbol $\supset$, as

$$B \supset A \tag{1.14}$$

A moment of thought will lead one to the conclusion that if A is a proper subset of B, it is also a subset of B. The converse is not necessarily true. A set S is a subset of itself, but it can't be a proper subset of itself.

Example 1.8

Given the following sets:

$$A = \{a, b, c\}$$
$$B = \{a, b, c\} \tag{1.15}$$
$$C = \{a, b, c, d\}$$

one can make the following statements:

$$A = B$$
$$B \supseteq A$$
$$A \subseteq B \tag{1.16}$$
$$C \supset A \text{ and } C \supseteq A$$
$$A \subset C \text{ and } A \subseteq C$$

The last two statements also apply with B substituted for A.

There are many sets that are deemed important in mathematics and appear so frequently that particular symbols are reserved for them. The symbols and the sets they represent are shown in Table 1.1.

TABLE 1.1

Symbols Reserved to Denote Important Sets

Symbol	Name of Set
Ø	Null Set
Z	Integers
N	Natural Numbers
Q	Rational Numbers
R	Real Numbers
C	Complex Numbers
U	Universal Set

1.2 Set Operations

Having established what a set is, the question turns to what can one do with sets? There are several basic operations over sets.

1.2.1 Union

In simple mathematics, there is the concept of addition. Sets have a similar concept, called *union*. The union of two sets produces a set that contains all objects that are members of either set. The notation for union is ∪.

Example 1.9

Given the following sets:

$$A = \{a, b, c\}$$

(1.17)

$$B = \{c, d, e\}$$

the union of A and B is

$$A \cup B = \{a, b, c, d, e\}$$

(1.18)

It should be noted that while c appears in both A and B, it only appears once in C since something can't be a member in a set more than once. The Venn diagram illustrating this example is shown in Figure 1.4. What it shows is two sets, A and B, with their borders identified with lines and the union of the two sets illustrated using shading to identify the resultant set.

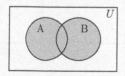

FIGURE 1.4
Venn diagram for A ∪ B.

An obvious property of union is that a set A is a subset of the set produced by taking a union of A with another set B:

$$A \subseteq (A \cup B) \tag{1.19}$$

It makes sense since the union is guaranteed to contain A since all members of A will be in the union of it with any other set.

The union operator is communitive, meaning that the order of the operands is not significant:

$$A \cup B = B \cup A \tag{1.20}$$

This makes sense since the order of elements within a set does not matter. The set is given identity by its members, not the order in which members appear within the set.

The union operator is associative, meaning that it doesn't matter how one groups together union operations. Thus,

$$A \cup (B \cup C) = (A \cup B) \cup C \tag{1.21}$$

It follows that in an expression such as the above, that the parenthesis could be removed without changing the meaning of the expression. That is,

$$A \cup (B \cup C) = A \cup B \cup C \tag{1.22}$$

If one thinks about the associative property carefully, one realizes that since the order of elements within a set is not important, then the order in which elements are added to set isn't important. A Venn diagram illustrating the associative property appears in Figure 1.5.

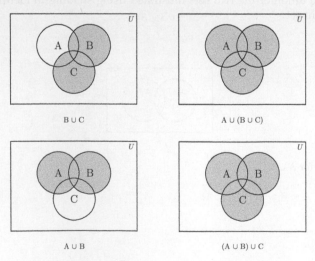

FIGURE 1.5
Associative property of union.

Two important properties of union are that the union of a set with itself produces the set itself and the union of a set with the null set produces the set itself. Thus, one can write

$$A \cup A = A$$

$$A \cup \varnothing = A$$

(1.23)

This property can be used to simplify expressions involving unions of sets.

One final property of interest relates subsets and unions. This property is

$$A \subseteq B \text{ if and only if } A \cup B = B$$

(1.24)

If A had an element that wasn't in B, then the union of A and B wouldn't produce B. Thus, A must only contain members that are in B. This also implies that A could be the null set. One could use this to conclude that the null set is a subset of any given set.

1.2.2 Intersection

Another operation that is closely related to union is *intersection*, denoted with the symbol $\cap$. Whereas a union is the set which results from combining two sets, the intersection is a set that results from taking only members that are members of both sets.

Example 1.10

Given the following sets:

$$A = \{a, b, c\}$$

$$B = \{c, d, e\}$$

(1.25)

then the intersection of the two sets is

$$A \cap B = \{c\}$$

(1.26)

A Venn diagram illustrating this example is shown in Figure 1.6.

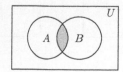

FIGURE 1.6
$A \cap B$.

Disjoint sets occur when two sets have no members in common except for the empty set. In this case,

$$A \cap B = \{\} \tag{1.27}$$

The intersection operator is communitive, as shown below:

$$A \cap B = B \cap A \tag{1.28}$$

This makes sense since intersection produces a set that contains only the members that are common between the two sets and the order of elements within the set does not matter. The intersection operator is associative, which means

$$(A \cap B) \cap C = A \cap (B \cap C) = A \cap B \cap C \tag{1.29}$$

Again, this is an obvious consequence in that any intersection is going to produce a set that contains only those elements that are common among all of the sets. This is illustrated in Figure 1.7.

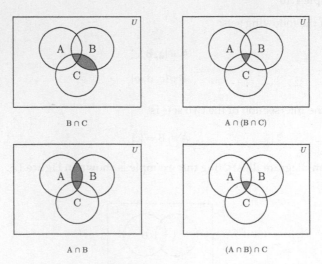

FIGURE 1.7
Associative property of intersection.

1.2.3 Set Difference

Set difference, also known as the *relative complement*, can be viewed as the opposite of union. While it has similarities with subtraction, set difference is denoted by "\" rather than a "−". If A and B are sets, the set difference, A\B, produces a set containing those elements that are in A and are not in B.

Example 1.11

Consider the following sets:

$$A = \{a, b, c, d\}$$
$$B = \{c, d, e, f\} \tag{1.30}$$
$$C = \{1, 2, 3\}$$

Taking the difference of A with B, A with C, and B with A leads to

$$A \backslash B \ = \ \{a, b, \cancel{c}, \cancel{d}\} = \{a, b\}$$
$$A \backslash C = \{a, b, c, d\} \tag{1.31}$$
$$B \backslash A = \{\cancel{c}, \cancel{d}, e, f\} = \{e, f\}$$

where the members being removed as a result of the difference is shown with the strikethrough. As one can see from the example, A\B is not equal to B\A.

There are three trivial special cases that should be obvious from the definition:

$$A \backslash A = \varnothing$$
$$\varnothing \backslash A = \varnothing \tag{1.32}$$
$$A \backslash \varnothing = A$$

First, taking the set difference of a set with itself removes from the set all elements that are in it, thus producing an empty set. Second, since an empty set does not contain any elements, then there are no elements that need to be removed from it regardless of what is in it, thereby producing an empty set. Finally, removing nothing (the empty set has no elements) from a set leaves the original set unchanged.

There are a number of identities dealing with set difference, union, and intersection that are rather important. The first of these is

$$C \backslash (A \cap B) = (C \backslash A) \cup (C \backslash B) \tag{1.33}$$

This might not make immediate sense but examining the Venn diagram in Figure 1.8 will clarify things significantly.

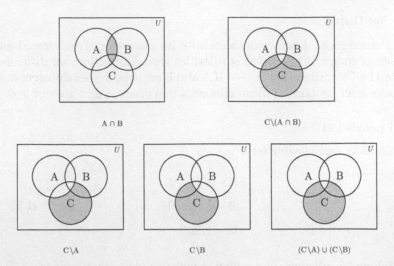

FIGURE 1.8
$C\backslash(A \cap B) = (C\backslash A) \cup (C\backslash B)$.

On the left-hand side, $A \cap B$ finds all the elements common to A and B. Those are then removed from C. On the right-hand side, $C\backslash A$ removes all elements of A from C, including those common to both A and B, while leaving elements which are unique to B in the set. The second expression $(C\backslash B)$ removes all elements of B from C, including those common to A but will leave those that are unique to A in the set. So, both expressions remove those elements that are common to A and B. The union of those two results leave C without the elements that are common to A and B, in other words $C\backslash(A \cap B)$.

Another identity that is important is

$$C\backslash(A \cup B) = (C\backslash A) \cap (C\backslash B) \tag{1.34}$$

This is easily visualized using the Venn diagrams in Figure 1.9. $(C\backslash A)$ produces a set which contains all elements of C except those that are in A. $(C\backslash B)$ produces a set which contains all elements of C except those that are in B. The intersection of $(C\backslash A)$ and $(C\backslash B)$ will contain only those elements that remain in C after elements that are either in A or B are removed. This is the same as removing $(A \cup B)$ from C.

A less obvious identity is

$$C\backslash(B\backslash A) = (C \cap A) \cup (C\backslash B) \tag{1.35}$$

Again, this is easily understood using Venn diagrams as shown in Figure 1.10. $(B\backslash A)$ removes all of A from B. $C\backslash(B\backslash A)$ removes from C all

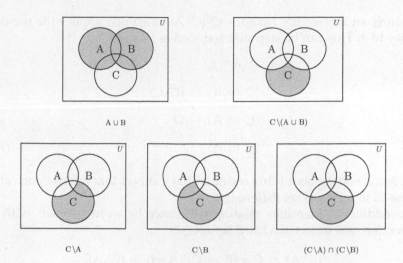

FIGURE 1.9
C\(A ∪ B) = (C\A) ∩ (C\B).

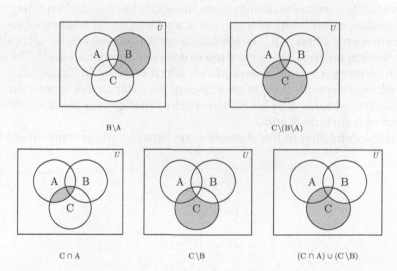

FIGURE 1.10
C\(B\A) = (C ∩ A) ∪ (C\B).

those elements of B that are not also in A. Having removed A from (B\A) will allow any element that is in A to remain in C. So, C will have (C ∩ A) as members. It will also have to delete any members of B that remain after members of A have been removed from it. All the other members of B have to be removed from C, thus (C\B). The result then becomes the union of the two subsets, (C ∩ A) ∩ (C\B).

There is an interesting case for C\(C\A) when one applies the previous identity to it. This can be stepwise reduced as follows:

$$C \backslash (C \backslash A)$$

$$(C \cap A) \cup (C \backslash C)$$

$$(C \cap A) \cup \varnothing$$

$$(C \cap A) \tag{1.36}$$

What is interesting about this result is that it shows that intersection can be expressed in terms of set difference.

Two additional identities relating difference between two sets with the intersection and union of a third set are

$$(B \backslash A) \cap C = (B \cap C) \backslash A = B \cap (C \backslash A)$$
$$(B \backslash A) \cup C = (B \cup C) \backslash (A \backslash C) \tag{1.37}$$

It is a valuable exercise to demonstrate these identities using Venn diagrams.

The relative complement of a set A\B is a set that contains all members of A which are not members of B. The *absolute complement* of a set A is the set of all elements that are not in A. It assumes the existence of a universal set U. The absolute complement A is thus given by U\A, which is shown in Figure 1.11.

The absolute complement is of sufficient importance that a notation for it has been introduced, $A^c = U \backslash A$. Alternative notations for the absolute complement of A include $\bar{A}$ and A'.

From the definition of the absolute complement, certain complement laws hold:

$$\varnothing^c = U$$

$$U^c = \varnothing$$

$$A \cup A^c = U \tag{1.38}$$

$$A \cap A^c = \varnothing$$

$U \backslash A$

FIGURE 1.11
U\A.

The first expression states that the absolute complement of a set that contains nothing is the universal set. The second states that the absolute complement of the universal set that contains everything is the empty set that contains nothing. The union of everything in A with everything that is not in A produces the universal set U. The intersection of a set with its complement is empty. According to the last two equations, if A is a proper subset of U, then A and A^c creates a *partition* of the set U, {A, A^c}.

The *double complement* law, also known as *involution*, states that the complement of the complement of a set produces the original set, as

$$(A^c)^c = A \qquad (1.39)$$

Think of it this way, if we take the complement of a set, the result is a set that contains everything that wasn't in the original set. If the complement is then taken, the result will contain only those members that weren't in the first complement, thus producing the original set.

De Morgan's laws allow the expression of unions and intersections purely in terms of each other via complements. They can be expressed in English as follows:

- The complement of the union of two sets is the same as the intersection of their complements.
- The complement of the intersection of two sets is the same as the union of their complements.

Using set notation, De Morgan's laws are expressed as

$$(A \cup B)^c = A^c \cap B^c$$
$$(A \cap B)^c = A^c \cup B^c \qquad (1.40)$$

These are easy to understand using Venn diagrams as illustrated in Figures 1.12 and 1.13.

There is one final form of difference—the *symmetric difference*, also known as the *disjunctive difference*. The symmetric difference, A $\triangle$ B, of sets A and B produces a set that contains elements that are unique to A and unique to B. The symmetric difference is illustrated in the Venn diagram in Figure 1.14.

This difference can be expressed as

$$A \triangle B = (A \backslash B) \cup (B \backslash A) \qquad (1.41)$$

or alternatively,

$$A \triangle B = (A \cup B) \backslash (A \cap B) \qquad (1.42)$$

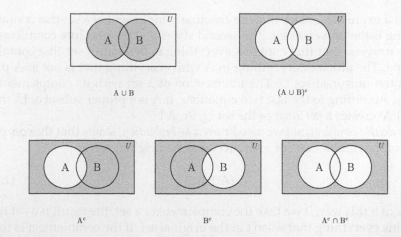

FIGURE 1.12
$(A \cup B)^c = A^c \cap B^c$.

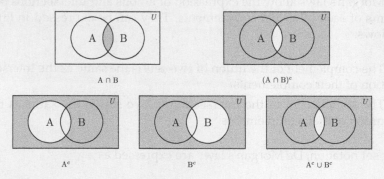

FIGURE 1.13
$(A \cap B)^c = A^c \cup B^c$.

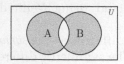

FIGURE 1.14
Venn diagram for symmetric difference.

The first form basically says that $A \triangle B$ is equal to everything in A that is not in B combined with everything in B that is not in A. The second form says that $A \triangle B$ is equal to the union of A with B less the elements that are in common between A and B.

One can use the commutative properties of union to show that the disjunctive union is commutative:

$$A \vartriangle B = (A \backslash B) \cup (B \backslash A) \qquad \text{Definition}$$
$$(B \backslash A) \cup (A \backslash B) \qquad \text{Union is communitive} \qquad (1.43)$$
$$B \vartriangle A \qquad \text{Definition}$$

The associative property of union can also be used to show that the disjunctive union is associative:

$$(A \vartriangle B) \vartriangle C = A \vartriangle (B \vartriangle C) \qquad (1.44)$$

Every set is its own inverse, meaning

$$A \vartriangle A = \varnothing \qquad (1.45)$$

That is, the symmetric difference of a set with itself is empty. The empty set is neutral:

$$A \vartriangle \varnothing = A \qquad (1.46)$$

This is easy to prove from the definition and the properties of difference and union:

$$A \vartriangle \varnothing = (A \backslash \varnothing) \cup (\varnothing \backslash A) \qquad \text{Definition}$$
$$= A \cup (\varnothing \backslash A) \qquad \text{Property } A \backslash \varnothing = A$$
$$= A \cup \varnothing \qquad \text{Property } \varnothing \backslash A = \varnothing \qquad (1.47)$$
$$= A \qquad \text{Property } A \cup \varnothing = A$$

1.2.4 Power Set

The power set of a set A is a set that contains:

All possible subsets of A as its members
No other members

The power set of A is denoted as $\wp(A)$. Notice the power set is expressed in terms of subsets rather than proper subsets.

Example 1.12

Consider the following set:

$$A = \{1,2,3\} \qquad (1.48)$$

The power set of A is the set made up of all possible subsets of A,

$$\wp(A) = \{\{\}, \{1\}, \{2\}, \{3\}, \{1,2\}, \{1,3\}, \{2,3\}, \{1,2,3\}\} \qquad (1.49)$$

Notice that it contains the empty set and the set itself.

1.3 Ordered Pairs

A set does not impose order on its elements. That is, $\{a, b\}$ is the same set as $\{b, a\}$. There are some cases where order is important. Sets can be used to define an ordered pair, written as $<a, b>$, in which a is the first member and b is the second member. The definition is

$$<a, b> = \{\{a\}, \{a, b\}\} \qquad (1.50)$$

The first member is taken to be the element that appears in the singleton and the second member is the one that is a member of the other set, but not of $\{a\}$. Looking at the definition, it is obvious that $<a, b> \neq <b, a>$ since $\{\{a\}, \{a, b\}\} \neq \{\{b\}, \{b, a\}\}$ unless $a = b$.

The definition can be extended to support ordered triplets as

$$<a, b, c> = <<a, b>, c> \qquad (1.51)$$

One can easily establish how higher-order ordered sets can be defined. An ordered set is often referred to as a "tuple." Note that when referred to as a tuple, an ordered set is denoted using a notation with the elements of the set enclosed in simple parentheses—for example (a, b, c)—with the elements present in the intended order.

Example 1.13

A question that is occasionally asked is, how can one have an ordered pair of the form $<a, a>$, since a appears more than once in it? Using the definition for an ordered pair, it can be shown that $<a, a> = \{\{a\}\}$. The steps are trivial:

$$<a, a> = \{\{a\}, \{a, a\}\}$$

$$= \{\{a\}, \{a\}\} \qquad (1.52)$$

$$= \{\{a\}\}$$

It may seem wrong, but $<a, a> = \{\{a\}\}$.

Consider the case in which there are two sets, A and B. One can form an ordered pair using an element from A as the first member and an element from B as the second member. That is, we can have <a, b> with a $\in$ A and b $\in$ B.

The *Cartesian product* of A and B, denoted as A $\times$ B, is the set of all ordered pairs that can be constructed with elements of A as the first member and elements of B as the second member. This can be defined as

$$A \times B = \{<a, b> | a \in A \text{ and } b \in B\}$$

It must be understood that if A or B is the empty set, then A $\times$ B is the empty set. Interestingly, the set produced by taking the Cartesian product is not ordered, although the elements within it are ordered pairs.

Example 1.14

Consider the following two sets:

$$A = \{a, b\}$$
$$B = \{1, 2, 3\} \tag{1.53}$$

The cross product of A and B, A $\times$ B, is

$$\{<a, 1>, <a, 2>, <a, 3>, <b, 1>, <b, 2>, <b, 3>\} \tag{1.54}$$

and B $\times$ A is

$$\{<1, a>, <1, b>, <2, a>, <2, b>, <3, a>, <3, b>\} \tag{1.55}$$

while A $\times$ A is

$$\{<a, a>, <a, b>, <b, a>, <b, b>\} \tag{1.56}$$

One can readily observe that A $\times$ B is not equal to B $\times$ A.

1.4 Relations

In much the same way as people have an informal understanding of sets, people have an informal understanding of relations. Parent-of is a relation between a person and a person who is that person's child. We also understand that the parent-of relation does not exist between every pair of people, only specific people. We recognize relations through transitive verbs.

Transitive verbs typically have a subject and an object, with the verb representing an activity that is transferred from the subject to the object. Thus, the word *bit* establishes a bites relation between a dog and a mailman in the sentence, "The dog bit the mailman."

The objects within one set may exist in relations with objects in a second set. This is often denoted with notation aRb (infix notation) or Rab (prefix notation), if the relation R holds between the specific objects a and b. There is a direction implied here, namely that the relation holds from a to b. This is an ordered pair, <a, b>.

The following condition,

$$\text{if } a \in A \text{ and } b \in B \text{, then a relation R can be written as } R \subseteq A \times B$$

can be used to describe all ordered pairs that exist in the relation between elements of A and elements of B. The expression $R \subseteq A \times B$ denotes a relation from A to B. The projection of R on A, namely the set of elements in A that appears as the first element in the ordered pairs of R, is called the *domain*. The projection of R on B, namely the set of elements in B that are the second elements in the ordered pairs of R, is called the *range*.

Example 1.15

Consider sets A and B along with a relation R,

$$A = \{a_1, a_2\}$$

$$B = \{b_1, b_2\} \tag{1.57}$$

$$R = \{<a_1, b_1>, <a_1, b_2>, <a_2, b_2>\}$$

The domain of R is $\{a_1, a_2\}$ and the range of R is $\{b_1, b_2\}$ (Figure 1.15).

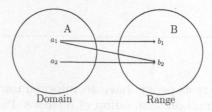

FIGURE 1.15
Domain A and range B of relation R.

Example 1.16

Consider sets A and B along with a relation R,

$$A = \{a_1, a_2\}$$
$$B = \{b_1, b_2, b_3\} \tag{1.58}$$
$$R = \{<a_1, b_1>, <a_1, b_3>, <a_2, b_1>\}$$

The domain of R is $\{a_1, a_2\}$ and the range of R is $\{b_1, b_3\}$. Note that the range doesn't include b_2 since it does not appear as the second element in any of the ordered pairs in R.

One can identify $A \times B$ as defining the universal set of which only some members are in R. Since not every possible element from $A \times B$ might not be in R, it is then possible to take the complement of R. The complement of a relation can be defined as follows:

$$R^c = (A \times B) \backslash R \tag{1.59}$$

This is the set of all ordered pairs in $A \times B$ that are not in R.

Example 1.17

Consider sets A and B along with a relation R,

$$A = \{a_1, a_2\}$$
$$B = \{b_1, b_2\} \tag{1.60}$$
$$R = \{<a_1, b_1>, <a_1, b_2>, <a_2, b_1>\}$$

The complement of the relation R would be:

$$R^c = A \times B \backslash R$$
$$= \{<a_1, b_1>, <a_1, b_2>, <a_2, b_1>, <a_2, b_2>\} \backslash \tag{1.61}$$
$$\{<a_1, b_1>, <a_1, b_2>, <a_2, b_1>\}$$
$$= \{<a_2, b_2>\}$$

The complement of R is not to be confused with the inverse R^{-1} of the relation, R. The inverse is the set of ordered pairs in R with the first and second elements reversed. Using the previous example,

$$R^{-1} = \{<b_1, a_1>, <b_2, a_1>, <b_1, a_2>\} \tag{1.62}$$

A relation and its inverse relation must satisfy the following condition,

$$\text{If } R \subseteq A \times B, \text{ then } R^{-1} \subseteq B \times A \qquad (1.63)$$

The complement, R^{-1}, is R with the domain and range reversed.

1.5 Functions

While one often thinks of a function as a computation, in reality it is a special case of a relation. A relation, R, from A to B is a function if and only if the following two conditions hold true:

1. Each element in the domain is paired with just one element in the range
2. The domain R is equal to A

These conditions are much more restrictive than one might initially consider.

Example 1.18

Given two sets A and B,

$$A = \{a_1, a_2\}$$
$$B = \{b_1, b_2\} \qquad (1.64)$$

then the following relations are functions:

$$F_1 = \{<a_1, b_1>, <a_2, b_1>\}$$
$$F_2 = \{<a_1, b_2>, <a_2, b_1>\} \qquad (1.65)$$
$$F_3 = \{<a_1, b_1>, <a_2, b_2>\}$$

These are all functions because the domain of each is the same as A and because each element in the domain is paired with one and only one element in the range. The following relations are not functions:

$$R_1 = \{<a_1, b_1>, <a_1, b_2>, <a_2, b_1>\}$$
$$R_2 = \{<a_1, b_2>\} \qquad (1.66)$$

R_1 is not a function because a_1 is paired with both b_1 and b_2, thus violating the first condition. R_2 is not a function because the domain of R_2, which is $\{a_1\}$ is not equal to A.

With a function being a special case of a relation, the notation of F: A → B is used to identify that it is a function and that it is a function from A to B.

A function can also be viewed as a map, mapping, transformation, or correspondence. There is a tendency to consider a function to be some computed entity according to some mathematical equation, but the fact is that a lookup table is a perfectly fine example of a function.

Example 1.19

The ASCII character set, partially shown in Table 1.2, is a mapping of an integer value to a character. The domain is the set of integers from 0 to 127. The range is over a set of characters, not all of which are printable.

$$A = \{0, 1 \ldots 127\}$$

$$B = \{NUL, SOH, STX, RTX \ldots, A, \ldots Z, \ldots DEL\}$$

With R

$$R = \{<0, NUL>, <1, SOH>, <2, STX>, \ldots, <127, DEL>\}$$

The table is a function since every element in the domain appears once and is paired with one value from the range.

It is also the case that elements in the domain are called the *arguments* of the function and their corresponding elements in the range are called the values. This can be denoted using the representation:

$$F(a) = b \tag{1.67}$$

where a ∈ A in which A is the domain and b ∈ B in which B is the range. The name of the function proceeds the argument that is enclosed in

TABLE 1.2

ASCII Character Set

Dec	Char	Dec	Char	Dec	Char	Dec	Char
0	NUL	32	Space	64	@	96	`
1	SOH	33	!	65	A	97	a
2	STX	34	"	66	B	98	b
3	ETX	35	#	67	C	99	c
...							
29	GS	61	=	93	]	125	}
30	RS	62	>	94	^	126	~
31	US	63	?	95	_	127	DEL

parenthesis and the corresponding value to the right of the equal sign. This is the form of representing a function that most people recognize as a function.

Example 1.20

Given two sets A and B, and a function F,

$$A = \{a_1, a_2\}$$

$$B = \{b_1, b_2\} \tag{1.68}$$

$$F = \{<a_1, b_2>, <a_2, b_1>\}$$

It can be said that F takes on the value of b_2 at argument a_1 and the value of b_1 at argument a_2. This can be written as

$$F(a_1) = b_2$$

$$F(a_2) = b_1 \tag{1.69}$$

If a function is a special case of a relation, then why do we tend to associate functions with computations? That's because a computation is one way of identifying the value of a function for a given argument. It is also a very succinct way to express the relation between two very large sets. The function

$$F = \{<x, y> | y = x * x\}$$

where x and y are integers is the same as

$$y = F(x) = x * x$$

Again, it must be reiterated, that not all functions use a computational means of correlating argument and value. The use of tables that represent key-value pairs is exceptionally common in computer science.

1.5.1 Partial Functions

Given that a relation, R, from A to B is a function if and only if the following two conditions hold true:

1. Each element in the domain is paired with just one element in the range
2. The domain R is equal to A

Then what about the cases where one of the conditions is not met? Specifically, what happens when the relation satisfies the first, but not the second? Such a relation is referred to as a *partial function*.

Example 1.21

Consider the inverse of an integer:

$$R(x) = 1/x \tag{1.70}$$

This clearly satisfies the first condition since there is only one value for any given integer except zero. The problem is that it doesn't satisfy the second condition since 0 is an integer, but 1/0 is undefined. Thus $R(x)$ is a partial function.

1.5.2 Into/Onto

As stated previously, a function is a mapping of elements within A to elements within B. Functions from A to B are described as a mapping of elements from A *into* B. The mapping into B does not necessarily have to produce every value within B, only that for every element in A, there is some element in B. If the range of the function is equal to B, then the mapping is said to be *onto*. This means that for any element in B, there is one and only one element in A that maps to it. The difference between into and onto is shown in Figure 1.16.

A mapping is said to be one-to-one, if for every element in B, there is only one element in A that maps to it. A mapping is said to be many-to-one if more than one element in A maps to an element in B.

A function that is onto and one-to-one is called a one-to-one correspondence. It should be noted that the inverse of a one-to-one correspondence is also a function, but the inverse of a many-to-one function is not.

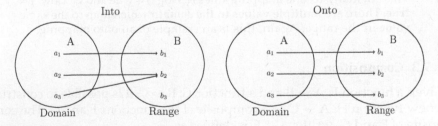

FIGURE 1.16
Into and onto illustrated.

Example 1.22

Consider the sets

$$A = \{1, 2, 3\}$$

$$B = \{4, 5, 6\}$$

A function F: A $\rightarrow$ B of the form F(x) = 6 is a perfectly fine function that maps all elements of A into B, yet only one element of B is in the range.

Example 1.23

Consider the case where

$$A = \{1, 2, 3\}$$

$$B = \{4, 5, 6\} \tag{1.71}$$

$$F: A \rightarrow B \text{ is of the from } F(x) = x + 3$$

This is a one-to-one mapping since F(1) = 4, F(2) = 5, and F(3) = 6. It is also an onto mapping since the range of the function is equal to B.

Example 1.24

Consider the case where

$$A = \{\text{"Cat", "Dog", "Cow"}\}$$

$$B = \{\text{true, false}\} \tag{1.72}$$

$$F(x) = \text{true if and only if the word contains an "o"}$$

This is a many-to-one mapping since F("Dog") = true and F("Cow") = true. There are multiple values in the domain which map to the same value in the range. Again, this is an example of an onto mapping.

1.5.3 Composition

Given a function F: A $\rightarrow$ B and a function G: B $\rightarrow$ C, it is possible to construct a new function H: A $\rightarrow$ C as a composite of the functions F and G. The composite of F and G, written G $\circ$ F, is defined as

$$G \circ F =_{def} \{<x, z> | \text{for y, } <x, y> \in F \text{ and } <y, z> \in G\} \tag{1.73}$$

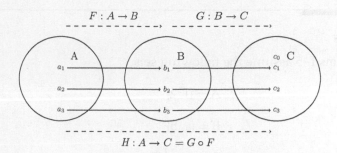

FIGURE 1.17
Function composition.

The notation $G \circ F$ might seem contrary to expectation, but according to the definition the composition can be written as: $G(F(x)) = z$. Viewed in that manner, the order makes sense (Figure 1.17).

Compositions do not have to be formed from one-to-one, one-to-many, or onto mappings. All that is required is that G and F are both functions. F and G can be disjoint, meaning that there is no y such that $<x, y>$ and $<y, z>$ exists, so that the composition is the empty set.

A function $F: A \rightarrow A$ such that $F = \{<x, x> | x \in A\}$ is called the *identity* function, written as id_A. Given the definition of composition, it should be noted that a composite of a function, F, with the appropriate identity function, id_A, is F. That is,

$$id_A \circ F = F \tag{1.74}$$

The concept of identity for composite functions has interesting consequences when considered with respect to functions that are one-to-one correspondences. In a one-to-one correspondence $F: A \rightarrow B$, there is an inverse function $F^{-1}: B \rightarrow A$. Using the definition of composition and that for inverse, it can be shown that the following equations must hold true:

$$F \circ F^{-1} = id_A$$

$$F^{-1} \circ F = id_B \tag{1.75}$$

Composition is a more general concept than function. Since functions are a special case of relations everything about composition for functions can also apply to relations. For any relation R of the form, $R \subseteq A \times B$ and relation $S \subseteq B \times C$, the composite, $S \circ R$, is the relation defined by

$$S \circ R = \{<x, z> | <x, y> \in R \text{ and } <y, z> \in S\} \tag{1.76}$$

Exercises

For Problems 1–5 assume the following sets:

$$A = \{1, 2\}$$

$$B = \{a, b\}$$

$$C = \{1, 2, 3\}$$

$$D = \{a, b, c\}$$

$$E = \{1, 2, a, b\}$$

$$U = \{1, 2, 3, a, b, c\}$$

1. Evaluate the following for true or false:
   a. $1 \in A$
   b. $b \in A$
   c. $c \notin A$
   d. $1 \notin E$
   e. $b \in D$

2. Compute the following:
   a. $A \cup B$
   b. $A \cup C$
   c. $A \cap D$
   d. $B \cap A$
   e. $A \cap E$
   f. $(D \cap E) \cup A$
   g. $(D \cap E) \cup B$
   h. $(D \cap E) \cap B$

3. Evaluate the following for true or false:
   a. $A \subseteq A$
   b. $A \subset A$
   c. $A \subseteq C$
   d. $E \supset C$
   e. $B \subset E$
   f. $(A \cup B) \subset E$
   g. $(A \cup B) = D$

4. Compute the following:
   a. E\A
   b. E\B
   c. C\D
   d. A\E
   e. E\E
   f. (E\A) ∪ (E\B)
   g. U\A
   h. A\U
   i. A Δ B
   j. ℘(C)

5. Compute the following:
   a. B × A
   b. D × D
   c. (C × D)\(A × B)

For Problems 6–8 assume:

$$A = \{1, 2, 3\}$$

$$B = \{a, b, c\}$$

$$C = \{1, 2\}$$

$$D = \{a, b\}$$

$$R1 = \{<1,b>, <1, c>, <2, a>, <3, b>\}$$

$$R2 = \{<1,c>, <2, b>, <3, a>\}$$

$$R3 = \{<1,1>, <1, 3>, <2, 1>, <2, 2>\}$$

$$R4 = \{<1,c>, <2, c>, <3, a>\}$$

6. Identify the domain and range of the following:
   a. R1
   b. R2
   c. R3

7. Compute the following:
   a. A × B\R1
   b. C × D\R3
   c. A × B\R1

8. Answer the following questions:
    a. Which of R1, R2, R3, and R4 are functions?
    b. With $R3 \subseteq C \times A$, what is the complement of R3?
    c. What is the inverse of R3?
    d. Is R2 an into or onto mapping?
    e. Is R2 a one-to-one or a one to many mapping?
    f. Of the functions among the four relations, which one has an inverse function?

2

Finite State Machine

How are the functions of a coin-operated vending machine specified? How is the current value of coins calculated in the vending machine, and how is it determined when the machine can dispense a product? How does it know to dispense a particular product? Or how is the behavior of a traffic light described? How does an elevator respond to presses of a button? Is pressing the up button twice actually going to bring the elevator sooner?

At first pass, it seems easy enough to answer any of these questions using simple language. One puts coins into a vending machine until a sufficient value of coins has been inserted and then it can dispense a product. A button is then pressed that tells it which product to dispense. The problem with that approach is that kind of description isn't robust enough to program against. There are the other details that are important, such as knowing when enough money has been inserted, how to make change, and handling a return coin press correctly. In fact, text isn't a very good way to specify these things. What happens is that a set of disconnected statements about the machine is generated, but the overall functioning of the machine is not specified.

Things tend to exhibit three kinds of behavior: static behavior, continuous behavior, and state-based behavior. Static behavior is constant and can be specified using a simple function. For example, the mass of an object is an attribute of the object with a constant value. Continuous behavior encompasses smoothly changing behavior based on its history, such as a weather vane that always points in the direction in which the wind is blowing. State behavior occurs when an object has different discrete behaviors, the manifestation of which depends on its state. A good example is the overhead light in a car which has three states: *always on, always off,* and *door-controlled.* In the *always on* state, the light will remain on regardless of whether the car door is open or closed. In the *always off* state, the light will remain off even when the door is opened. In the *door-controlled* state, the light will turn on when the door is opened and turn off when the door is closed.

While static and continuous behaviors are relatively easy to specify, a more rigorous way to specify state behaviors is required. One such formalism is the finite state machine (FSM), or as it is alternatively known, finite state automaton (FSA), state machine (SM), and finite automaton (FA). What is a finite state machine? It is a mathematical model of computation. It is an abstract machine that can be in one and only one state at a given time, has a special state called the start state, and can change from one state to another via a transition activated because of some input.

The kinds of systems that finite state machines are good at describing include *recognizers*, *classifiers*, *transducers*, and *generators*. A recognizer, also called an acceptor or sequence detector, produces a binary output indicating whether some sequence of inputs is accepted, such as one might want in validating a password. A classifier is like a recognizer, but rather than producing a binary output it can produce one of multiple output values if a sequence of inputs is accepted, such as one might want in a self-diagnostic system that can not only identify that a system is in an errored state, but also identify the error. A transducer, usually used for control systems or computational linguistics, generates output based on the input and/or the state of the system using actions. A generator, also known as a sequencer, produces an output sequence based on inputs.

One of the powers of using a finite state machine to model certain types of systems is that it allows one to focus on issues one at a time. One can concentrate on individual states independently of the transitions that connect state to state. One can give full attention to the transitions without worrying about what happens inside a state. One can focus on any actions that are associated with states and/or transitions.

2.1 Key Concepts of Finite State Machines

As stated previously, a finite state machine is a mathematical model of computation describing an abstract machine. At any given time, the machine can only be in exactly one of a finite number of states. The FSM can change from that state to another (undergo a transition) in response to some external input and/or when a condition is satisfied. The transition can incorporate a set of actions to be executed when the machine goes from one state to the next.

At the heart of FSMs is the concept of a state. We can say that a state is a distinguishable, disjoint, orthogonal, and persistent mode of being. To say that a state is distinguishable means that it accepts inputs and exhibits behaviors that make it different than other states. To say that the states of a machine are disjoint, means that an object can only be in one state in a time and must be in exactly one state at all times. Being orthogonal means that states do not overlap other states. Persistent means that the object is always in a state and can remain in a given state for a period of time.

While a machine, at any given time, is described by the current state within which it exists, the machine as whole is described by the set of states within which the machine *can* exist, the state within which the machine exists upon creation, the transitions that the machine can execute, and the inputs which the machine recognizes. A classic formal definition of a state diagram for an FSM or FA is a directed graph described by the tuple (Q, Σ, δ, q_0) where

Q is a finite set of states.

Σ is a finite collection of inputs.

δ is a set of directed transitions from one state to another.

q_0 is the start state $q_0 \in Q$.

Typically, members of the set of transitions, δ, are members of the relation, $Q \times \Sigma \rightarrow Q$ or the relation $Q \times \Sigma \rightarrow \wp(Q)$ with $\wp(Q)$ a subset of Q. The case where δ is given by $Q \times \Sigma \rightarrow Q$ is called a *deterministic finite automata* (DFA), since for every state and input there is at most one transition to another state. The case where δ is given by $Q \times \Sigma \rightarrow \wp(Q)$ is called a *nondeterministic finite automata* (NFA) since for some states and a given input there are possible transitions to multiple states.

The mathematical model described by a given tuple can be used to guide implementation of a state machine, yet it isn't so good for understanding the machine or its behavior. What one really wants is to be able to get an understanding of the entire machine as a single gestalt. One way to achieve that goal is to have a single figure, a simple state diagram, that presents the individual parts as elements of the whole. The directed graph has states as nodes and transitions as arcs with the arch labeled by the input which activates the transition.

In a simple state diagram, a state (a node) is denoted by a circle with a designator symbol or word written inside it to label the state, illustrated to the left in Figure 2.1. All state machines have a start state which is the state a machine enters upon creation. There's nothing special about a start state except that it is the initial state of the machine. The start state is identified by an arrow pointing into it that does not originate in another state as shown to the right in Figure 2.1. In some cases, there is a terminal state, or final state, which brings the state machine to a halt (there is no transition that leads from that state).

Since a transition (a directed arc) describes how a state machine goes from one state to another, it is shown by an arrow connecting an originating state to a next state with the arrowhead pointing into the next state. The arrow is labeled with the input that causes the transition. This is shown in Figure 2.2.

Simple State   Initial State

FIGURE 2.1
Representation of states in a simple state diagram.

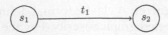

FIGURE 2.2
Documenting transitions in a simple state diagram.

Example 2.1

Consider a deterministic state machine described by the tuple (Q, Σ, δ, q_0) where

$$Q = \{s_0, s_1, s_2, s_3, s_4\}$$

$$\Sigma = \{i_1, i_2, i_3\}$$

$$\delta = \{<<s_0, i_1>, s_1>, <<s_1, i_2>, s_2>, <<s_2, i_2>, s_3>, <<s_2, i_3>, s_0>, <<s_0, i_3>, s_4>\}$$

$$q_0 = s_0$$

While the above description provides everything that one needs to understand about the state machine, it isn't obvious what the machine does. One can ask if there is a sequence of transitions that takes the machine from its start state back to its start state. The answer is yes, but finding the sequence is not easy. A simple state diagram for the machine allows one to understand it with much less effort and answering the question becomes almost trivial (Figure 2.3).

The way in which the directed transitions are specified in the previous example, is rather difficult to read and understand. An alternative way of expressing the directed transitions is to use a transition table. The states are represented in rows with inputs as columns. The cell of a given state and input identifies the state to which it transitions. If there is no transition

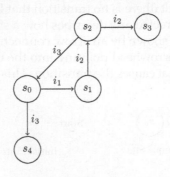

FIGURE 2.3
Simple state machine diagram.

TABLE 2.1

State Transition Table

State	i_1	i_2	i_3
s_0	s_1	–	s_4
s_1	–	s_2	–
s_2	–	s_3	s_0
s_3	–	–	–
s_4	–	–	–

for a given state and input, a simple dash is placed to denote that nothing happens. The table or the previous example would be expressed as shown in Table 2.1. From this transition table, it is obvious that states s_3 and s_4 are final states since there are no transitions out of them.

2.2 Accepting States

Recognizing that a particular set of inputs has been entered or that particular sequences can be categorized as significant are two uses of finite state machines. One way of accomplishing this is to mark some states as accepting. Accepting states are regular states which are only significant when the state machine consumes all of the input and ends up in that state. In that case, the machine is said to accept the input stream. If on the other hand, the state machines consumes all input and ends up in a state which is not an accepting state, it is said that the state machine rejects that input stream.

A formal definition of a state diagram for a finite state machine or finite automaton with accepting states is a directed graph described by the tuple (Q, Σ, δ, q_0, F) where:

Q is a finite set of states.

Σ is a finite collection of inputs.

δ is a set of transitions from one state to another as caused by the input.

q_0 is the start state $q_0 \in Q$.

F is a set of accepting states.

The mathematical model is what can be used to guide implementation of a state machine. There can be DFA and NFA versions of state machines with accepting states.

Example 2.2

Let's consider the example of a DFA for determining if some binary string represents an odd number. This is easily determined if the binary string ends in 1. Taking into account the state where nothing has been read in, the FSA for this machine is

Q: {start, even, odd}

Σ: {0,1}

δ: see Table 2.2

q_0: start

F: {odd}

The diagram for this FSA is shown in Figure 2.4. From this diagram, it is easy to tell that the binary string 1010 will end in the even state and hence won't be in the accepting state odd. However, the binary string 1011 will end in the odd state which is recognized as an accepting state.

TABLE 2.2

State Transition Table for DFA

State	0	1
Start	Even	Odd
Even	Even	Odd
Odd	Even	Odd

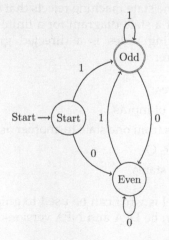

FIGURE 2.4
DFA for determining if binary string is odd or even.

Example 2.3

Let's revisit the previous example but use an NFA to model the situation. The mathematical description is similar to the description for a DFA. Rather than start, odd, and even, this machine has a start state, an odd state, and an accumulator state (acc) which consumes 0s and 1s. For a given input of 1, there are multiple choices in states into which the machine can transition.

Q: {start, acc, odd}

Σ: {0,1}

δ: see Table 2.3

q_0: start

F: {odd}

As can be seen from the state transition table, the odd state is a final state. The state diagram for this FSA is shown in Figure 2.5.

In this case, it is not quite so clear what happens while consuming the binary string. This machine has a choice of two possible transitions in accumulator state for a 1. However, if the sequence of FSA transitions leads into the odd state, then there are no possible transitions out of it. If there are still digits of the binary string to be read, then that sequences is invalid.

Let's consider the possible sequences of transitions for the string 1010. In sequence 1, the first transition to odd leads to a dead end, so that is

TABLE 2.3

State Transition Table for NFA

State	0	1
start	acc	acc, odd
acc	acc	acc, odd
odd	–	–

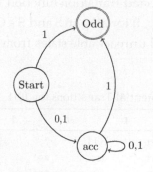

FIGURE 2.5
NFA to determine if binary string is odd or even.

TABLE 2.4

Sequence of Potential Transitions for 1010

Sequence		1	0	1	0
1	start	odd	?		
2	start	acc	acc	odd	?
3	start	acc	acc	acc	acc

not a possible sequence. In sequence 2, the transition to odd after reading the third digit prevents the machine from being able to process any other digits. It is only in sequence 3, that the machine is able to process all of the digits of the string (Table 2.4).

Processing the binary string 1011 is more complicated in that it ends with two possible sequences after consuming all of the digits of the input string (Table 2.5). Sequences 1 and 2 both end in a final state before the binary string has been fully consumed. That makes those two sequences unacceptable. Sequences 3 and 4 end up in different final states of odd and acc respectively. This raises the question, is the string odd? In an NFA, if any completed sequence of transitions ends in an accepting state, then the outcome is accepted. Thus, 1011 is an odd binary number because a sequence which consumes the full string ends in accepting state, odd. Many people find this behavior somewhat unsettling.

Given a state machine, M, with set of states Q_N, initial state q_0, inputs Σ, transitions δ_N, and accepting states F_N, we will produce an equivalent state machine, M', which is deterministic. The procedure is simple:

1. Generate the power set over Q_N as $Q'_D = P(Q_N)$.
2. Create a state for every element in Q'_D.
3. Set $\{q_0\}$ as the initial state.
4. Any state which includes at least one accepting state in the original machine is set to be accepting.
5. Generate a new directed transition function, δ_D, using the formula.
    a. $\delta_{D'}(S, i) = \cup \delta_N(p, i)$ for all p in S and S $\in$ Q'_D.
6. Remove useless and unreachable states from Q'_D to generate Q_D.

TABLE 2.5

Sequence of Potential Transitions for 1011

Sequence		1	0	1	1
1	start	odd	?		
2	start	acc	acc	odd	?
3	start	acc	acc	acc	odd
4	start	acc	acc	acc	acc

This appears complicated, but it is actually a very simple procedure as shown in the following example.

Example 2.4

This example shows how to convert the NFA of the previous example into a DFA. We start with the following:

Q_N: {start, acc, odd}

Σ: {0,1}

δ_N: see Table 2.3

q_0: start

F: {odd}

The first step is to compute the set Q'_D as the power set over Q_N. Since there are three states, the power set will contain $2^3 - 1$ states (which excludes the empty set). The result is as follows:

Q'_D = {{start}, {acc}, {odd}, {start, acc}, {start, odd}, {acc, odd}, {start, acc, odd}}

The next step is to convert each member of the powerset into a state:

$$s_0 = \{start\}$$

$$s_1 = \{acc\}$$

$$s_2 = \{odd\}$$

$$s_3 = \{start, acc\}$$

$$s_4 = \{start, odd\}$$

$$s_5 = \{acc, odd\},$$

$$s_6 = \{start, acc, odd\}$$

This initial state q_{N0} is set to s_0 since it is the state comprised of the start state only. The next step is to construct an intermediary set containing the accepting states. Each state in Q'_D, which includes an accepting state within it is marked as an accepting state. Thus,

$$F'_N = \{s_2, s_4, s_5, s_6\}$$

The state diagram, excluding transitions, is shown in Figure 2.6.

We can now construct a new state transition table. To give an example for two of the states:

$$\delta_D(s_3, 0) = \delta_N(start, 0) \cup \delta_N(acc, 0) = \{acc\} \cup \{acc\} = \{acc\} = s_1$$

FIGURE 2.6
Intermediate state diagram without transitions.

$$\delta_D(s_3, 1) = \delta_N(\text{start}, 1) \cup \delta_N(\text{acc}, 1) = \{\text{acc}, \text{odd}\} \cup \{\text{acc}, \text{odd}\} = \{\text{acc}, \text{odd}\} = s_5$$

$$\delta_D(s_5, 0) = \delta_N(\text{acc}, 0) \cup \delta_N(\text{odd}, 0) = \{\text{acc}\} \cup \{\} = \{\text{acc}\} = s_1$$

$$\delta_D(s_5, 1) = \delta_N(\text{acc}, 1) \cup \delta_N(\text{odd}, 1) = \{\text{acc}, \text{odd}\} \cup \{\} = \{\text{acc}, \text{odd}\} = s_5$$

The transition table will have seven states as rows and two inputs as columns. This table is shown in Table 2.6. The state diagram including all of the states and state transitions is shown in Figure 2.7.

Looking at the transition table and the state diagram, it is obvious that states S_2, S_3, S_4, and S_6 have no transitions into them, which means that they can never be reached. These states can be removed from Q'_D to produce Q_D. The state machine can now be described as (Table 2.7):

Q_D: $\{s_0, s_1, s_5\}$

Σ: $\{0, 1\}$

δ_D: see Table 2.2

q_0: s_0

F_D: $\{s_5\}$

TABLE 2.6

Transition Diagram for Constructed States

State	0	1
s_0	$\{\text{acc}\} = s_1$	$\{\text{acc}, \text{odd}\} = s_5$
s_1	$\{\text{acc}\} = s_1$	$\{\text{acc}, \text{odd}\} = s_5$
s_2	–	–
s_3	$\{\text{acc}\} = s_1$	$\{\text{acc}, \text{odd}\} = s_5$
s_4	$\{\text{acc}\} = s_1$	$\{\text{acc}, \text{odd}\} = s_5$
s_5	$\{\text{acc}\} = s_1$	$\{\text{acc}, \text{odd}\} = s_5$
s_6	$\{\text{acc}\} = s_1$	$\{\text{acc}, \text{odd}\} = s_5$

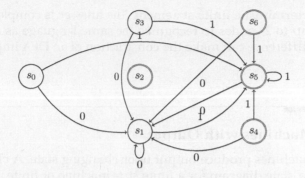

FIGURE 2.7
Full state transition diagram for constructed states.

TABLE 2.7

State Transition Table for Result of
Transforming Nondeterministic FDA
to DFA

State	0	1
Start	s_1	s_5
Even	s_1	s_5
Odd	s_1	s_5

The state diagram for this is shown in Figure 2.8. As can be seen, this
is structurally identical to the previous example of a deterministic state
machine for determining if a binary string is odd. The only difference is
the labels of the states.

Every DFA can be treated as a special case of an NFA. As will be shown
shortly, every NFA can be converted into a DFA. Considering the difference
in possible state sequences, one might wonder why one would ever bother

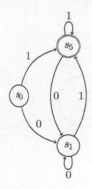

FIGURE 2.8
State diagram for transformed nondeterministic FDA.

with a nondeterministic finite statement. The answer is complexity. A DFA can require up to 2^n nodes to recognize the same language as an NFA of n nodes. This difference can make the construction of an DFA impractical.

2.3 State Machines with Output

Some state machines produce output upon changing state. A classic formal definition of a state diagram for a finite state machine or finite automaton is a directed graph described by the tuple $(Q, \Sigma, Z, \delta, q_0, F)$ where:

Q is a finite set of states.

Σ is a finite collection of inputs.

Z is a finite collection of outputs.

δ is set of directed transitions.

q_0 is the start state $q_0 \in Q$.

F is a set of accepting states.

The mathematical model is what can be used to guide implementation of a state machine.

Documenting output is a little tricky since an output can be associated with the transition or the state. If it is associated with the transition, it is identified under the input causing the event with a line separating them. If it is associated with the state, it is identified under the label for the state with a line separating the two. This is shown in Figure 2.9.

The state machines of the previous examples did not incorporate outputs, that is, $Z = \{\}$. There are three major models for how outputs are incorporated in state machines: Mealy, Moore, and Harel.

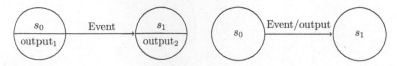

FIGURE 2.9
Documenting output.

2.3.1 Mealy

A Mealy machine is a 6-tuple $(Q, q_0, \Sigma, Z, \delta, G)$ consisting of the following:

Q is a finite set of states.

q_0 is the start state $q_0 \in Q$.

Σ is a finite collection of inputs.

Z is a finite collection of outputs.

δ is a set of directed transitions $Q \times \Sigma \to Q$.

G is a set of output functions $Q \times \Sigma \to Z$.

In some formulations, the transition and output functions are coalesced into a single function:

$$\delta: Q \times \Sigma \to Q \times Z$$

In a practical sense, the output is associated with the transition from a given state for a given input. The simple state diagram for a Mealy machine illustrates this by associating the output with the transition as shown in Figure 2.10. This machine is an edge detector that outputs 1 whenever it detects a change between two adjacent digits. It does not report the first digit as an edge.

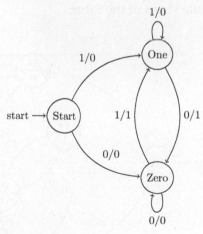

FIGURE 2.10
Mealy simple state diagram.

2.3.2 Moore

A Moore machine is a 6-tuple $(Q, q_0, \Sigma, Z, \delta, G)$ consisting of the following:

Q is a finite set of states.

q_0 is the start state $q_0 \in Q$.

Σ is a finite collection of inputs.

Z is a finite collection of outputs.

δ is a set of directed transitions $Q \times \Sigma \to Q$.

G is a set of output functions $Q \to Z$.

This looks very much like a Mealy machine except that the output is associated with the state rather than the transition. The simple state diagram for a Moore machine illustrates this by associating the output with the transition as shown in Figure 2.11. The machine shown here also implements an edge detector. It is a lot more complicated than the Mealy machine due to the fact that the states have to track the previous and the current digit.

2.3.3 Harel (UML)

While Mealy and Moore machines are capable of modeling a lot of systems, problems start to arise when the system becomes complex. For example, the state of a car can be described in terms of the states of all of its parts (lights, engine, brakes, windshield wipers, etc.). A Mealy or Moore machine would have to create composite states of the form:

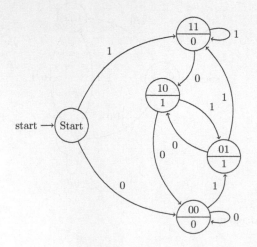

FIGURE 2.11
State diagram for a Moore machine.

```
lights-on_engine-on_brakes-off_wipers-on ...
lights-off_engine-on_brakes-off_wipers-on ...
lights-on_engine-off_brakes-off_wipers-on ...
```

The problem is that one has an explosion of states. Adding a new part that has two states doubles the number of states of the original machine. There's also a problem when the machine incorporates some internal attribute that can slowly change, but that the states of the machine change only when certain thresholds on the attribute are met.

The Harel model avoids the problems of both Mealy and Moore machines by allowing complex machines to be constructed of simpler machines. A Harel machine supports

- Nested states
- Actions and activities
- Guards
- History
- Concurrency
- Broadcast transitions
- Orthogonal components

It supports all of this at the expense of not having a formal definition in the same way as a Mealy or Moore machine. Nested states enable modeling a ceiling fan that has two major states, on and off, along with three nested states for slow, medium, and fast fan speeds. Actions and activities represent different kinds of outputs: actions being a discrete act and an activity being an ongoing act. Guards allow constraints to be placed on transitions based on attribute values. History allows one to model a CD player in an automobile that is able to resume playing the appropriate track after the CD has been suspended due to shutting off the car. Concurrency enables modeling systems that can have more than one activity going on at a time. Broadcast transitions allow for generating events that are sent to other state machines as a result of a transition. Finally, the use of orthogonal components allows for modeling a car with its myriad subsystems that all function, more or less, independently.

A Harel machine is often diagrammed using a UML (unified modeling language) state diagram. The simple UML state diagram captures simple states along with transitions and activities. A UML state diagram for an initial state, two simple states, and a final state is shown in Figure 2.12.

The figure shows the start state as a simple filled-in circle that transitions directly to State1 without a trigger being necessary. It shows the final state as a filled-in circle surrounded by a slightly larger circle. Each state has an internal transitions compartment that contains the actions to be performed on entry and exit of the state, along with the activity to be performed while

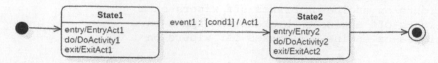

FIGURE 2.12
Simple state diagram.

in the state. It is not mandatory to specify entry and exit actions and/or do activities within a state since they may or may not exist. The transitions are shown with the trigger that causes the transition, a guard condition, and an activity that is performed during the transition. The guard condition and the activity associated with the transitions are optional in specification.

Assuming the event occurs that initializes the transition from Start to State1, the sequence in which the actions and activities occur is significant. On entry to State1, the entry action, EntryAct1, is performed. The doActivity1 is begun upon the completion of EntryAct1. The doActivity1 will continue until the trigger, event1, occurs (assuming it satisfies the guard condition) and initiates the transition to State2. The doActivity1 will be ended, the exit action, ExitAct1, will be performed, and then transition will be fired at which time the action, Act1, that is associated with the transition will be executed. Once Act1 has completed, the machine will enter State2. On entry, EntryAct2 will execute, and once it has completed the activity, doActivity2 will begin.

The basic sequence on receiving an event that causes a transition is as follows:

1. The activity identified by the do in the original state will end.
2. The exit action of the original state will execute.
3. The transition begins with the execution of the action associated with the transition.
4. When the transition action completes the next state is entered.
5. On entry to the next state, the entry action is performed.
6. Once the entry action has completed, the activity identified by the do will begin.

If any step depends on an activity that is not specified, then that step is skipped. A Mealy machine would only have a specification for the action associated with step 3. A Moore machine would only have an action associated with step 5. Thus, even when dealing with simple states, the Harel machine can capture a significantly more complex specification of behavior than a Mealy or Moore model.

The arrival of an event that satisfies a guard condition does not necessarily trigger an external transition (one in which the machine leaves the original state and enters a destination state). An internal transition only executes an action appropriate for that event. An internal transition compartment contains a list of internal transitions, where each item has the form as described for trigger. Each event name may appear more than once per state if the guard conditions are different. The event parameters and the guard conditions are optional. If the event has parameters, they can be used in the expression through the current event variable.

Example 2.5

Figure 2.13 illustrates how to represent a ceiling fan that can be off or on and that has low, medium, and high speeds when on. Transitioning from each state is a result of pulling the chain on the ceiling fan. Note in the example how an initial state element inside the on state is used to select which internal state the fan will enter.

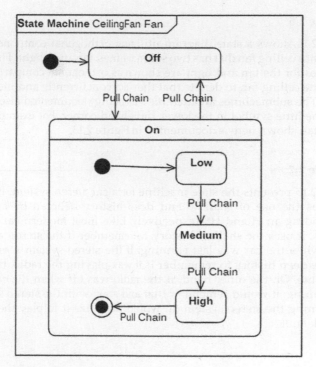

FIGURE 2.13
State machine for the multispeed fan of a ceiling fan.

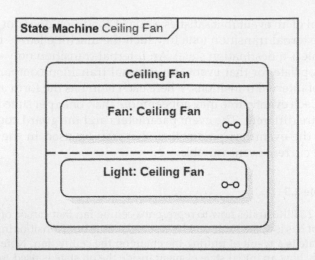

FIGURE 2.14
State diagram for a ceiling fan with light and fan submachines.

Example 2.6

Figure 2.14 shows a state diagram utilizing orthogonal components to represent a ceiling fan that has two submachines: fan and a light. The state machines for the fan and light are shown as orthogonal compartments inside the ceiling fan to denote that they act concurrently and independently. The submachines are identified as being documented elsewhere using the little symbol in the lower right-hand corner. For example, the ceiling fan shown here is documented in Figure 2.13.

Example 2.7

Figure 2.15 presents the state machine for a car stereo system. It demonstrates the use of shallow and deep history denoted by a circle surrounding an H and H*, respectively. Like most modern car stereo systems, it uses the shallow history to remember if the stereo was on or off when the car was last running. If the stereo system was on, it uses the deep history to remember if it was playing the radio, the CD, or satellite. On the other hand, if the radio was off when the car was last running, it would remember that and start with the stereo system off. Turning the stereo system on would initialize it to play the radio as the default.

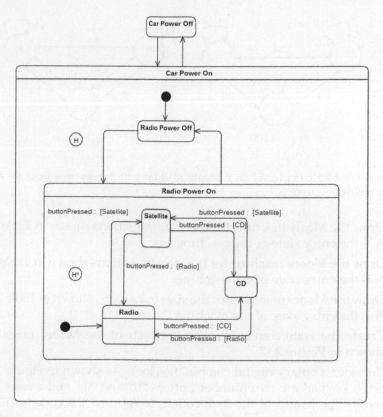

FIGURE 2.15
Car radio sound system.

Exercises

1. Create the state transition table for the Mealy machine shown in Figure 2.16a. What is the output sequence for an input sequence 0011101?

2. Create the state transition table for the Mealy machine shown in Figure 2.16b. What is the output sequence for an input sequence abaabba?

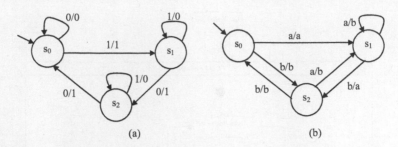

FIGURE 2.16
Two three-state Mealy machines. (a) A Mealy machine with inputs from {0, 1}. (b) A Mealy machine with inputs from {a, b}.

3. Draw the Mealy machine for the state transitions shown in Table 2.8. A is the entry state of the machine.

4. Draw the Moore machine for the state transitions shown in Table 2.9. A is the entry state of the machine.

5. Draw the Moore machine for the state transitions shown in Table 2.10. A is the entry state of the machine.

6. Create the state transition table for each of the Moore machines shown in Figure 2.17.

7. Consider a binary digital combination lock. As shown in Figure 2.18, on its keypad are two number buttons "0" and "1," and a reset button R. The length of the lock code is 6. Assume the code is 110101. Whenever the lock detects such an input sequence, it is unlocked. For example, any one of the following binary sequences will unlock it:

110101

0110101

011110110101

1110001101110101

TABLE 2.8

The State Transition Table for Problem 3

Present State	Input	Next State	Output
A	a	A	b
	b	B	a
B	a	C	–
	b	B	a
C	a	A	a
	b	D	a
D	a	A	b
	b	D	–

TABLE 2.9

The State Transition Table for Problem 4

Present State	Input	Next State	Output
A	a	A	–
	b	B	
B	a	C	b
	b	B	
C	a	A	a
	b	D	
D	a	A	a
	b	D	

TABLE 2.10

The State Transition Table of Problem 5

Present State	Input	Next State	Output
A	0	A	0
	1	B	
B	0	E	0
	1	C	
C	0	D	1
	1	C	
D	0	A	1
	1	F	
E	0	A	0
	1	F	
F	0	E	1
	1	C	

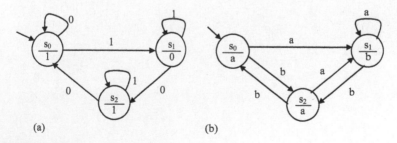

(a)   (b)

FIGURE 2.17

Two three-state Moore machines. (a) A Moore machine with inputs from {0, 1}. (b) A Moore machine with inputs from {a, b}.

FIGURE 2.18
Keypad of a binary digital combination lock.

At any time, a press on the reset button will take the lock to its initial state. The lock enters its final state when the sequence of the code is detected. Design a DFA and draw the state transition diagram for this lock.

8. Develop a Harel state machine to specify the dynamic behavior of a door that can be opened and closed. When it is closed it can be locked and unlocked. Notice that you can open or close a door only if the doorway is cleared.

   a. Use only simple states.

   b. Use nested states to model the behavior of the door when it is closed. The nested states are the two internal states: *unlocked* and *locked*.

9. Draw the state machine for a simple battery charger that can charge two batteries in parallel. The charger has three modes: idle, discharging, and charging. The charger starts charging the batteries when a button is pushed. However, before each battery can be charged, it must be discharged. When each battery is discharged, it sends a signal to the charger to change the mode to charging. When a battery is charged, it sends notifies to the charger. When both batteries are charged, the charger returns to the idle mode.

3

Regular Expressions and Languages

A lot of people think they understand language, but it is a superficial understanding. Part of the reason for that is that people are born into a language, learn to speak it, read it, and write it. They often study a grammar for the language in terms of nouns, verbs, adjectives, and adverbs. That's perfectly fine for a natural language like English, Spanish, or German. This is knowledge of a single language, but not an understanding of what it means to be a language.

A deeper understanding of language starts with the alphabet. An alphabet is any finite set of symbols. This includes symbols that aren't letters, as one might think. Typical examples of symbols include, letters, digits, and punctuation. Some important alphabets include binary {0,1}, ASCII character set, and Unicode.

3.1 Strings and Languages

Often in programming languages, a string is denoted within double quotes, but that is not what is meant here. A string over an alphabet is a finite sequence of symbols drawn from that alphabet. Words and sentence are often used as synonyms for string. The length of a string, s, usually written as $|s|$, is the number of occurrences of symbols in s. The empty string, ε, is the string of zero length.

A *prefix* of a string is any string that can be created by removing symbols from the end of the string. A proper prefix has at least one symbol removed from the end of the string. A *suffix* of a string is any string that can be created by removing symbols from the beginning of the string. A proper suffix has at least on symbol removed from the beginning of the string. A *substring* of a string is any string that can be created by removing symbols from the beginning and end of the string. A proper substring has at least one symbol removed from either the beginning or end of the string. A *subsequence* of a string is any string that can be generated from the string by removing one or more positions of s. This doesn't require that the positions removed are consecutive.

Example 3.1

Consider the string banana. The following statements can be made about it:

- |banana| = 6
- banana is a prefix of banana
- ba is a proper prefix of banana
- ana is a suffix of banana
- nan is a substring of banana
- bnn is a subsequence of banana

A *language*, L, is any countable set of strings over some fixed alphabet. All syntactically correct programs are languages. All natural languages (English, Spanish, Chinese) are languages as one would expect. However, no specific meaning must be attached to the strings in the language. A language can be infinite (although it must still be countable).

Let L be the set of lowercase letters, {a–z}. L can be thought of in two ways. First, L can be considered an alphabet consisting of lowercase letters. Second, L can be considered a language comprised of strings of length one over an alphabet given by the set of lowercase letters. These two views are equivalent.

The most important operations that are performed on languages are union, concatenation, and closure. Union is the normal union of set theory, which is described in Chapter 1. One can consider the language defined by strings of length one drawn from the set of uppercase and the set of lowercase letters to be the union of two languages. Formally, one can say,

$$A \cup B = \{s \,|\, s \text{ is in } A \text{ or } s \text{ is in } B\}$$

That is, if A = {a–z} and B = {A–Z}, then the union of A and B, A ∪ B, produces a new language, L = {a–z, A–Z}.

One can think of concatenation as a cross product. Concatenation is combining two strings by attaching one to the end of the other to create a new string. If a and b are strings, then the concatenation of a and b is a string, denoted as ab. Formally,

$$AB = \{ab \,|\, a \text{ is in } A \text{ and } b \text{ is in } B\}$$

The empty string is the identity under concatenation

$$\varepsilon s = s\varepsilon = s$$

If concatenation is viewed as a product, then "exponentiation" of strings could be defined as

$$s^0 = \varepsilon$$

$$s^i = s^{i-1} s$$

Exponentiation of strings can be performed an infinite number of times.

Example 3.2

Assume that L is the set of uppercase and lowercase letters, {a–z, A–Z} and that D is the set of digits, {0–9}. Then,

$$L \cup D = \{a–z, A–Z, 0–9\}$$

Given that a–z is 26 characters, A–Z is 26 characters, and 0–9 is 10 characters, the set L ∪ D is composed of 62 strings. Each string will contain either a letter or a digit.

Example 3.3

Assume that L is the set of uppercase and lowercase letters, {a–z, A–Z} and that D is the set of digits, {0–9}. Then, LD is the set of strings of length 2 that have a letter in the first place and a digit in the second place:

$$LD = \{a0, a1, …, z9, A0, A1, … Z0\}$$

Since there are 52 letters that are combined with 10 digits, the total size of $|LD| = 520$.

The *closure* of a Language L, denoted as L^*, is the set of strings obtained by concatenating L zero or more times. L^n is the result of concatenating L n times. The union of concatenating a language, L, an infinite number of times produces the *Kleene Closure*, L^*,

$$L^* = \cup^{\infty}_{i=0} L^i$$

Based on this definition, L^* includes the empty string, ε. The *Positive Closure*, L^+, is the Kleene Closure without the term L^0, that is, it does not contain the empty set as the initial set.

$$L^+ = \cup^{\infty}_{i=1} L^i$$

Let's see the differences between the Positive Closure and the Kleene Closure.

Example 3.4

Let's assume that the language A is the alphabet {a, b}. The Positive Closure of A is

$$A^+ = \{a, b, aa, ab, ba, bb, aaa, aab, abb, …\}$$

The Kleene Closure of A is

$$A^* = \{ε, a, b, aa, ab, ba, bb, aaa, aab, abb, …\}$$

By using the operations of union, concatenation, and closure, languages can be built from other languages.

3.2 Regular Expressions

A legal identifier in the Java programming language must start with letter, an underscore, or a Unicode currency character which can be followed by one or more letters and digits. If one wanted to describe the set of valid Java identifiers, one would use a language such as

$$L(L \cup D)^*$$

with L being the set of letters {A...Z, a...z} and D being the set of digits {0...9}. Unfortunately, this does generate a language that includes elements (keywords and literals) which are not allowed as identifiers in Java and doesn't include the identifiers starting with an underscore or Unicode currency character.

Identifiers can be described by giving names to sets of letters and digits and using the language operators—union, concatenation, and closure—to produce strings that can be part of a language. This process is so useful that a notation called *regular expressions* has been created to identify how the languages can be built using these operators by applying them to the symbols of some alphabet.

Suppose we have the alphabets, letter_ and digit, given by

$$letter\_ = \{a–z, A–Z, \_\}$$

$$digit = \{0–9\}$$

The language of identifiers in programs could be described by

$$letter\_ (letter\_|digit)^*$$

with | representing union, * meaning "zero or more occurrences of," and the juxtaposition of letter_ with the remainder of the expression meaning concatenation.

This notation is a regular expression. Regular expressions are built recursively out of smaller regular expressions using rules. Each regular expression r denotes a language L(r), which is also defined recursively from the languages denoted by r's subexpressions.

Two forms of rules apply, basis rules and induction rules. The basis rules define the starting points for regular expressions and are as follows:

- ε is a regular expression, and L(ε) is {ε}, the language whose sole member is the empty string.
- If a is a symbol in Σ, then a is a regular expression and L(a) = {a} is the language with one string, of length one containing a in its one position.

The induction rules are how we can build new languages out of other languages. There are four such rules:

1. (r) denotes the language L(r)
2. (r)|(s) denotes the language L(r) ∪ L(s)
3. (r) (s) denotes the language L(r)L(s)
4. (r)* denotes the language (L(r))*

The first rule says that an extra parenthesis can be wrapped around an expression without changing the language it denotes. The second rule says that a new language can be formed by joining two languages. The third rule states that a new language can be formed by concatenating two languages. The fourth rule states that a new language can be generated by taking the closure of the original language.

Unnecessary pairs of parentheses can be removed by adopting the convention on precedence of operators:

- The unary operator * has highest precedence and is left associative.
- Concatenation has second highest precedence and is left associative.
- | has lowest precedence and is left associative.

A language that can be defined by a regular expression is called a *regular set*. If two regular expressions r and s denote the same regular set, then they are equivalent, denoted as r = s.

Because the induction rules are based on set operations, there are algebraic laws for regular expressions. These laws are shown in Table 3.1.

The first two entries in Table 3.1 result from the fact that | represents union and set union is communitive and associative. The next two entries follow from treating concatenation like a cross product. The fifth entry is obvious, an empty string concatenated with a string leaves just the string. The last entry is the result of the definition of the Kleene Closure, $L^* = \cup^{\infty}_{i=0} L$. The last

TABLE 3.1

Algebraic Laws for Regular Expressions

Law	Description
r|s = s|r	| is communitive
r|(s|t) = (r|s)|t	| is associative
r(st) = (rs)t	Concatenation is associative
r(s|t) = rs | rt; (s|t)r = sr | tr	Concatenation is distributive over |
εs = sε = s	ε is the identity for concatenation
r* = (r|ε)*	ε is guaranteed in a closure
r** = r*	* Is idempotent

one follows from $L^{**} = \cup^{\infty}_{i=0} (\cup^{\infty}_{i=0} L)$ that produces only duplicate elements which, in set theory, are replaced by a single element.

Utilizing regular expressions resulting from algebraic operations over a language can become cumbersome. For example, if

$$L(a\text{-}z, A\text{-}z, \_) = a|b|c|...|z|A|B|C|...|Z|\_$$

$$L(0\text{-}9) = 0|1|2|3|4|5|6|7|8|9$$

then the regular expression for an identifier is

$$(a|b|c|...|z|A|B|C|...|Z|\_)((a|b|c|...|z|A|B|C|...|Z|\_)$$
$$| (0|1|2|3|4|5|6|7|8|9))^*$$

This is, at best, awkward and difficult to read. To simplify the notation, we can name a regular expression and use the name instead of the full regular expression. Given an alphabet of basic symbols Σ, a regular definition is a sequence of definitions of the form

$$d1 \rightarrow r1$$

$$d2 \rightarrow r2$$

$$...$$

$$di \rightarrow ri$$

where di is a new symbol, not in Σ, and is unique from other d's and each ri is a regular expression over the alphabet $\Sigma \cup \{d1, d2, d3, ... di\}$. The awkward expression for an identifier, as given above, is simplified using this convention. That is, the expression is simplified to

$$letter\_ \rightarrow a|b|c|...|z|A|B|C|...|Z|\_$$

$$digit \rightarrow 0|1|2|3|4|5|6|7|8|9$$

$$id \rightarrow letter\_ (letter\_|digit)^*$$

Example 3.5

An integer is a sequence of digits that can be proceeded by an optional sign. A sequence of digits is a digit followed by one or more optional digits. With this in mind, an integer can be represented using this notation as

$$digit \rightarrow 0|1|2|3|4|5|6|7|8|9$$

$$optionalSign \rightarrow (+|-|\varepsilon)$$

$$digits \rightarrow digit\ digit^*$$

$$integer \rightarrow optionalSign\ digits$$

This means of specifying an integer reveals several practices. The first two lines identify basic alphabets as a regular expression. The second line incorporates the empty string, ε, which makes the sign optional. The third line enforces the fact that an integer must contain at least one digit, but that additional digits are optional since digit* includes the ε string as an element. The final line pulls together the definition of an integer by pairing the optional sign with one or more digits.

Some regular expression elements are so common that it makes sense to introduce additional operators to represent them as a form of shorthand. These operators represent an extension to the core operators of regular expressions. One subexpression that is exceptionally common in regular expressions is the optional element $(x|\varepsilon)$. The unitary postfix operator ? is used to eliminate the need for the | and the parenthesis in expressions containing optional elements. Thus, $(x|\varepsilon)$ can be replaced with the simpler notation x?, which means zero or one occurrence of x.

In the previous example, an extra line was used to define digits so that an integer would contain at least one digit. This very common practice corresponds to the Positive Closure rather than the Kleene Closure. The unitary postfix operator + is introduced to represent the Positive Closure. Thus, we can use r^+ to represent one or more r's. It follows that

$$r^+ = rr^* = r^*r$$

This defines the relationship between the Kleene Closure and the Positive Closure.

The first two lines of the example for the regular expression for an integer shows the specification of the language formed from the alphabet. That is, it included the optional + and − signs along with the specification of the digits as

$$digit \rightarrow 0|1|2|3|4|5|6|7|8|9$$

We can incorporate some syntactic sugar to avoid having to type|between every character. This is done by enclosing the characters within square braces, [], as

$$digit \rightarrow [0123456789]$$

One should notice that there are no spaces in the above. If a space were present than that would mean the alphabet of digits would include the space character. Space is not a delimiter, but a character in regular expressions. Of course, the digits are a well-known sequence. Laziness, which is not always a bad trait, dictates that a shorthand notation for well-known sequences be introduced. This is done by introducing a – between the beginning of the sequence and the end of the sequence. Digit can be written in a much simpler fashion as

$$\text{digit} \rightarrow [0\text{–}9]$$

The expression [a–z] represents the regular expression over the language formed from lowercase letters. The expression [a–l] represents the regular expression over the language formed by the first 12 lowercase letters.

Example 3.6

The previous example for specifying an integer can be simplified one line at time. The specification is simplified to

$$\text{digit} \rightarrow [0\text{–}9]$$

$$\text{digits} \rightarrow \text{digit}^+$$

$$\text{sign} \rightarrow [+\text{-}]$$

$$\text{integer} \rightarrow \text{sign? digits}$$

Notice that there is a space between "sign?" and "digits." This is to serve as a separator between names for supporting regular expressions, which is different than having a space between characters of the alphabet.

Example 3.7

One of the truisms of programming is garbage in/garbage out. What this tells us is that validating input is one critical factor for ensuring that programs function correctly. Admittedly, there is a big difference between validating that a string follows a pattern and authenticating the value. Take, for example, a zip code. A zip code has five digits that are followed by an optional hyphen followed by four digits or just an optional four digits. The regular expression to validate that an input conforms to a zip code is simple:

$$\text{digits5} \rightarrow [0\text{–}9]\ [0\text{–}9]\ [0\text{–}9]\ [0\text{–}9]\ [0\text{–}9]$$

$$\text{separator} \rightarrow [\text{-}|]$$

$$\text{digits4} \rightarrow [0\text{--}9]\ [0\text{--}9]\ [0\text{--}9]\ [0\text{--}9]$$

$$\text{zipcode} \rightarrow \text{digits5 (separator digits4)?}$$

This regular expression will validate zip codes of the following forms:

12345

12345 6789

12345-6789

It is much more difficult to show that it is a real zip code for some location and that the submission matches a city and state. However, going to a database and accessing a record to ensure that the zip code is correct is expensive. In this case, the first five digits correspond to a real zip code, but the addition of the last four digits makes it an invalid zip code.

3.3 Lex

Given an approach for expressing patterns using regular expressions and that this is based on mathematics, can we take the patterns for all the needed tokens to build a piece of code that examines the input string and finds a prefix that is a lexeme matching one of the patterns? That is, can we automate the generation of code for a lexical analyzer? The answer is yes. The caveat is that it is an advanced topic. While this book is not a primer on writing compilers or interpreters, some of the mathematical tools necessary to address this problem will be introduced in Chapter 8.

Lex is one program of many that is used for generating lexical analyzers based on regular expressions. A lexical analyzer is a program that processes strings and returns tokens that are recognized within the input string. The token identifies the recognized substring and associates attributes with it.

Lex uses regular expressions to specify recognizable tokens of a language. It uses regular expressions to identify what it should be able to recognize and to associate with that string a variable value. Lex uses an extended version of regular expressions that includes application of additional set operations (such as complement) beyond union, concatenation, and closure. The regular expressions recognized by Lex are also recognized by libraries that are available to programmers for processing regular expressions.

Example 3.8

The following is a Lex specification for recognizing the tokens of a calculator. Lines 1–4 are used to specify what elements of the generated program, such as the headers, to include and a global variable, c, which represents the ASCII value of the input matched by the regular expression.

```
1. % {#
2.    include < stdio.h > #include "y.tab.h"
3.    int c;
4. %}
5. % %
6. "";
7. [a-z] {
8.    c = yytext[0];
9.    yylval.a = c - 'a';
10.     return (LETTER);
11. }
12. [0-9] {
13.    c = yytext[0];
14.    yylval.a = c - '0';
15.    return (DIGIT);
16. }
17. [^a-z0-9\b]
18. {
19.    c = yytext[0];
20.    return (c);
21. }
22. % %
```

Line 5 identifies that the specification of strings to be recognized begins.

Line 6 identifies that it will recognize a space, but will return nothing. Note that Lex explicity specifies the string containing a single space.

Lines 7–11 identify that it will recognize a lowercase letter and that it will set a global variable yylval.a to a value of yytext[0]—"a" and returns the token LETTER. The variable yytext[0] contains the substring recognized by the regular expression. The subtraction of the character "a" converts it from the ASCII decimal value for the character to the position of the letter in the alphabet.

Lines 12–16 identify that it will recognize a digit, returns the token DIGIT, and sets yylval.a = yytext[0]—"0." The operation yytext[0]—')' converts the ASCII representation of the character to the numerical value of the character.

Lines 17–21 identify that it will recognize anything other than a space (denoted by\b), a character, and digit. It will return whatever it matches (which for a calculator should be a character corresponding to a mathematical operator).

3.4 Grammar

Typically, discussions of grammar in computer science focus on compilers and programming languages. It is true that every programming language must have precise rules to prescribe the syntactic structure of well-formed programs. The rules must be precise so that the compiler can be built around them since a compiler is nothing more than a program which processes the source code of a programming language. Computer programs can't deal with ambiguity.

Yet, the concept of grammar applies to more than just compilers and programming languages. When two programs communicate with each other, one program sends a message to the other program. That message must be parsed and analyzed with respect to its syntactic structure in exactly the same way that a compiler must process a source code for a program written in some programming language. Thus, specifying messages also involves languages and grammars.

Syntactic structure describes how a well-formed program or message is constructed in terms of its basic building blocks. In a structured programming language, a program is made up of functions, a function is made from declarations and statements, a statement is made from expressions, and so on. A message may be composed of parts, with subparts, attribute-value pairs, and so on. Even saying this, it is missing the details necessary to construct a program to process and recognize a well-formed program. A more precise way of expressing the structure must be found.

We can use a grammar to specify a programming language or message in the same way that a language like English uses grammar rules to describe a well-formed English sentence. A grammar gives a precise, yet easy-to-understand, syntactic specification of a programming language or message. The syntax of programming language constructs can be specified by context-free grammars that are expressed using BNF (Backus-Naur form) notation.

There are many different classes of grammars. Some of these allow us to automatically construct an efficient parser that determines the structure of a source program from the grammar itself. The structure imparted to a language by a properly designed grammar is useful for translating source programs into correct object code and for detecting errors. It can also be used to parse messages passed between programs, such as a web server and a web browser.

Often, we want our text processing capabilities to evolve. In most cases, we must evolve them iteratively by adding now constructs to perform new tasks or convey more information. For example, C++ evolved out of C to add object-oriented programming into a structured programming language. XML has undergone several iterations in which new elements of the basic exchange format have been added.

A grammar allows a language to evolve and develop in a controlled and iterative fashion. These new constructs can be integrated more easily into an implementation that follows the grammatical structure of the language. An important class of grammar is the *context-free* grammars.

A context-free grammar consists of terminals, nonterminals, a start symbol, and productions. Terminals are the basic symbols from which strings (sentences) are formed. The term "token name" or simply "token" is a synonym for "terminal."

Terminals are usually the first components of the token's output by the lexical analyzer. A token is a pair consisting of a token name and an optional attribute value. The token name is an abstract symbol representing a kind of lexical unit, such as a keyword, or sequence of input characters denoting an identifier. A pattern is a description of the form that the lexemes of a token may take. A lexeme is a sequence of characters in the source that matches the pattern for a token and is identified by the lexical analyzer, such as a program generated by LEX, as an instance of that token.

Nonterminals are syntactic variables that are used in the productions describing the grammar. These variables are used to denote sets of strings defining the language generated by the grammar by imposing a hierarchical structure.

One nonterminal is distinguished as the start symbol. The set of strings that can be derived from the start symbol based on the productions denotes is the language generated by the grammar. It is a convention for productions that are activated by the start symbol to be listed first in the specification for the grammar.

3.4.1 Productions

A grammar is specified in terms of a set of *productions*. The productions of a grammar are what determine the structure of the language. They specify the way terminals and nonterminals can be combined to create well-formed strings that are recognized in the language. Each production consists of the following:

- A nonterminal called the head, or left side, of the production that defines the strings which are denoted by the head.

- A body, or right side, consisting of zero or more terminals and non-terminals that describe one or more ways in which well-formed strings can be constructed from the nonterminal at the head.

- The symbol → (alternatively ::=) denotes the relation that maps the head to the body. Since the arrow is not supported in situations where ASCII characters are used, a common substitute for it is ::=.

This chapter gives preference to ::= over the → only because online queries for the BNF grammar specification for programming languages and exchange formats tend to return documents that use the former over the latter.

When a grammar is expressed using BNF notation,

- A nonterminal is identified by enclosing the name of the nonterminal in angle brackets, <>.
- A terminal is identified by enclosing it within single or double quotes.

Thus, the terminal for the digit 5 would be "5" and the nonterminal for a statement in a program might be denoted as <statement>.

Example 3.9

This example presents the BNF specification of the grammar for the BNF notation. One thing that should be noted here is that lines 8–19 would normally be defined using regular expressions with the lexemes identifying literals and symbols and provided to a grammar processor using a lexical analyzer.

```
1.  < syntax > ::= < rule > | < rule > < syntax >
2.  < rule > ::= < opt - whitespace >"<"< rule - name
    >">"< opt - whitespace >"::="< opt - whitespace >
    < expression > < line - end >
3.  < opt - whitespace > ::="" < opt - whitespace > |""
4.  < expression > ::= < list > | < list > < opt -
    whitespace >"|" < opt - whitespace > < expression >
5.  < line - end > ::= < opt - whitespace > < EOL > | <
    line - end > < line - end >
6.  < list > ::= < term > | < term > < opt - whitespace >
    < list >
7.  < term > ::= < literal > |"<"< rule - name >">"
8.  < literal > ::='"'< text1 >'"'|"'"< text2 >"'"
9.  < text1 > ::=""| < character1 > < text1 >
10. < text2 > ::=""| < character2 > < text2 >
11. < character > ::= < letter > | < digit > | < symbol >
12. < letter > ::="A" |"B" |"C" |"D" |"E" |"F" |"G" |"H"
    |"I" |"J" |"K" |"L" |"M" |"N" |"O" |"P" |"Q" |"R" |"S"
    |"T" |"U" |"V" |"W" |"X" |"Y" |"Z" |"a" |"b" |"c" |"d"
    |"e" |"f" |"g" |"h" |"i" |"j" |"k" |"l" |"m" |"n" |"o"
    |"p" |"q" |"r" |"s" |"t" |"u" |"v" |"w" |"x" |"y" |"z"
13. < digit > ::="0" |"1" |"2" |"3" |"4" |"5" |"6" |"7"
    |"8" |"9"
14. < symbol > ::="|" |"" |"-" |"!" |"#" |"$" |"%" |"&"
    |"("|")" |"*" |"+" |"," |"-" |"." |"/" |":" |";"
    |">"|"="|"<"|"?" |"@" |"["|"\" |"]"
15. "|" ^" |" " |" `" |"{"|"}" |"~"
16. <character1>      ::= <character> |"'"
17. <character2>      ::= <character> |'"'
18. <rule-name>       ::= <letter> | <rule-name> <rule-char>
19. <rule-char>       ::= <letter> | <digit> |"-"
```

Example 3.10

Let's examine the grammar for an expression that would be part of a programming language. An expression consists of terms separated by + signs, a term consists of factors separated by * signs, and factors can be either a parenthesized expression or an identifier. The grammar for this is

```
<expression>::= <expression> "+" <term> | <term>
<term>::= <term> "*" <factor> | <factor>
<factor>::= "("<expression> ")" | id
```

The above example is a *left-recursive* grammar, or as it is usually referenced, it is an LR grammar. Left recursive means that the grammar contains productions in which the leftmost element of the body side is the same as the head.

A left-recursive grammar is suitable for bottom-up parsing, but not top-down. In top-down parsing, applying the first production goes recursive and will never end as shown below:

```
<expression> ::= <expression> "+" <term> | <term>
<expression> ::= (<expression> "+" <term> | <term>) "+"
<term> | <term>
<expression> ::= ((<expression> "+" <term> | <term>) "+"
<term> | <term>) "+" <term> |<term>
```

An alternative grammar for the same language can be constructed that is not left recursive by introducing intermediate abstractions. The following is a set of non-left-recursive productions for the language:

```
<expression> ::= <term> <expression1>
<expression1> ::= "+" <term> <expression1> | ""
<term> ::= <factor> <term1>
<term1> ::= "*" <factor> <term1> | ""
<factor> ::= "(" <expression> ")" | <identifier>
```

This would be used for building a top-down parser.

These two grammars are describing the same language. The only difference is the kind of parsers that can be built for the two forms. People tend to understand and construct left-recursive grammars when initially trying to define a language.

Example 3.11

Using a syntactic variable *stmt* and *expr* to denote statements, <statement>, and expressions, <expression>, respectively, the production specifies the structure of an if-then-else conditional statement:

```
stmt → if (expr) stmt else stmt
```

In a full specification for a programming language, other productions, such as the one used in the previous example, would define precisely what an *expr* is and what else a *stmt* can be.

3.4.2 Derivations

The notion of derivations is very helpful for discussing the order in which productions are applied during parsing. The purpose of this is to understand the kinds of statements that are recognized by the grammar. In terms of parsing a sentence in the grammar, the construction of a parse tree can be made precisely by taking a derivational view, in which productions are treated as rewriting rules.

Beginning with the start symbol, each rewriting step replaces a nonterminal by the body of one of its productions. This derivational view corresponds to the top-down construction of a parse tree, but the precision afforded by derivations will be especially helpful when bottom-up parsing is discussed.

Example 3.12

The grammar below defines simple arithmetic expressions:

```
expression := expression + term
expression := expression - term
expression := -expression
expression := term
term := term * term
term := term/term
term := id
```

In this grammar, the terminal symbols are: id +- */(and). The nonterminal symbols are expression and term. Expression is the start symbol.

Normally, one would develop the grammar using meaningful nonterminals, such as expression, in productions. However, since the purpose here is to understand the mechanics of using a grammar, less informative and meaningful nonterminals will be used. The grammar will be simplified using E for expression and T for term. Rather than write each production for the same nonterminal, a shorthand in which the set of expressions can be combined using an asterisk *or* the symbol | will be adopted. With these modifications in mind, the grammar above becomes:

$$E := E + T|E - T|-E|T$$

$$T := T * T|T/T|id$$

This is a compact representation of the preceding grammar.

A class of derivations known as "rightmost" derivations is applied by rewriting the rightmost nonterminal at each step. Using E := -E as an example, the production signifies that if E denotes an expression, then −E must also denote an expression. The replacement of a single E by −E will be described by writing

$$E \Rightarrow -E$$

which is read as "E derives −E." The production E ⇒ (E) can be applied to replace any instance of E in any string of grammar symbols by (E). Thus,

$$E * E \Rightarrow (E) * E$$

$$E * E \Rightarrow E * (E)$$

With a single E and repeatedly applying productions in any order, one can get a sequence of replacements. For example,

$$E \Rightarrow -E \Rightarrow -(E) \Rightarrow -(T) \Rightarrow -(id)$$

Such a sequence of replacements is a derivation of −(id) from E. This derivation is a proof that the string −(id) is one particular and valid instance of an expression according to the grammar.

For a general definition of derivation, consider a nonterminal A in the middle of a sequence of grammar symbols, as in $\alpha A \beta$, where α and β are arbitrary strings of grammar symbols.

Suppose $A := \gamma$ is a production. Then, we write

$$\alpha \, A \, \beta \Rightarrow \alpha \, \gamma \, \beta$$

When a sequence of derivation steps $a_1 \Rightarrow a_2 \Rightarrow \ldots \Rightarrow a_n$ rewrites a_1 to a_n, we say a_1 derives a_n.

The symbol ⇒ means "derives in one step." Often it is necessary to say that something "derives in zero or more steps." In much the same way as applied to regular expressions, the * can be used to represent zero or more and the symbol $* \Rightarrow$ is used to denote this. Thus,

$a * \Rightarrow a$, for any string a, and

If $a * \Rightarrow \beta$ and $\beta \Rightarrow \gamma$, then $a * \Rightarrow \gamma$.

Likewise, $+ \Rightarrow$ means "derives in one or more steps."

If $S * \Rightarrow a$, where S is the start symbol of a grammar G, we say that a is a *sentential* form of G. Note that a sentential form may contain both terminals and nonterminals. It can also be empty. A *sentence* of G is a sentential form with no nonterminals. The *language* generated by a grammar is its set of sentences. Thus, a string of terminals w is in L(G), the language generated by G, if and only if w is a sentence of G (or $S * \Rightarrow w$).

A language that can be generated by a grammar is said to be a *context-free language*. If two grammars generate the same language, the grammars are said to be *equivalent*.

Example 3.13

The string $-(id + id)$ is a sentence of our example grammar because there is a derivation that can produce it:

$$E \Rightarrow -E \Rightarrow -(E) \Rightarrow -(E + E) \Rightarrow -(T + E) \Rightarrow -(id + E) \Rightarrow -(id + T) \Rightarrow -(id + id)$$

The strings $E, -E, -(E), \ldots, -(id + id)$ are all sentential forms of this grammar. $E \overset{*}{\Rightarrow} -(id + id)$ indicates that $-(id + id)$ can be derived from E.

It is important to note that the derivation from the start to a given sentential form may not be unique. An alternative derivation for $E \overset{*}{\Rightarrow} -(id + id)$ is

$$E \Rightarrow -E \Rightarrow -(E) \Rightarrow -(E + E) \Rightarrow -(E + T) \Rightarrow -(E + id) \Rightarrow -(T + id) \Rightarrow -(id + id)$$

While everything is the same through the first three derivations, the forth derivation diverges.

At each step in a derivation, there are two choices to be made:

1. It is necessary to choose which nonterminal to replace.
2. Having made this choice, it is necessary to pick a production with that nonterminal as head.

To understand how parsers work, it is important to consider how derivations in which the nonterminal to be replaced at each step are chosen. There are two basic strategies: leftmost derivations and rightmost derivations. In *leftmost derivations*, the leftmost nonterminal in each sentential is always chosen. The notation $a \Rightarrow_{lm} \beta$, is used to denote that $a \Rightarrow \beta$ is a step in which the leftmost nonterminal in a is replaced. If $S \Rightarrow_{ln} a$, then a is a left-sentential form of the grammar at hand. In *rightmost derivations*, the rightmost nonterminal is always chosen. This is denoted by writing $a \Rightarrow_{rm} \beta$. Rightmost derivations are sometimes called canonical derivations.

3.4.3 Parse Trees

A parse tree is a graphical representation of a derivation that has the advantage that it filters out the order in which productions are applied to replace nonterminals. Each interior node of a parse tree represents the application of a production. The interior node is labeled with the nonterminal A in the head of the production; the children of the node are labeled, from left to right, by the symbols in the body of the production by which this A was replaced during the derivation. The leaves of a parse tree are labeled by nonterminals or terminal. They are read from "left to right" and constitute a sentential form. This is sometimes called the yield or frontier of the tree.

Example 3.14

The example of E * ⇒−(id + id) can be used to demonstrate how a parse tree can be constructed from a derivation. This is shown in Figure 3.1. While this expression can be derived in more than one manner, the final parse tree does not convey leftmost or rightmost derivation.

It is unfortunate but true that a grammar can produce more than one parse tree for some sentence. Such a grammar is said to be ambiguous. An ambiguous grammar is one that produces more than one leftmost derivation or more than one rightmost derivation for the same sentence.

Normally, it is desirable that the grammar be made unambiguous, since if it is not, then it not possible to uniquely determine which parse tree to select for a sentence. In some cases, it is convenient to use an ambiguous grammar and incorporate disambiguating rules to eliminate undesirable parse trees.

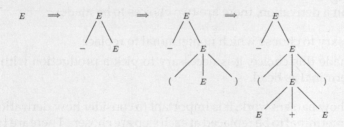

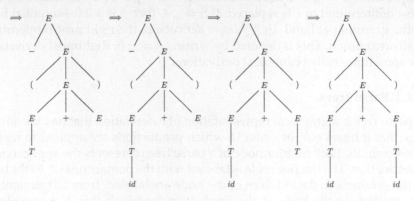

FIGURE 3.1
Parse tree for E * = >−(id + id).

Example 3.15

Consider the expression id + id * id. Based on the grammar that has been used so far, this can produce different parse trees as shown in Figure 3.2. The parse tree labeled a adds the last two numbers together before multiplying the result with the first number. The parse tree labeled b multiples the first two numbers together before adding the result to the last number. Mathematically, this produces very different results.

In this case, a disambiguation rule would be introduced that give precedence to multiplication, so that the proper parse would be selected as b rather than a. Sometimes an ambiguous grammar can be rewritten to eliminate the ambiguity.

Example 3.16

The infamous "dangling else" problem of programming is an example of a statement in which a grammar is ambiguous. The grammar for statements includes two forms for if statements:

stmt ⇒ if *expr* then *stmt* | if *expr* then *stmt* else *stmt* | ...

The problem of the dangling else arises when one has a statement of the form

If *expr* then if *expr* then *stmt* else *stmt*

There are two distinct parses for this statement illustrated by using parentheses to distinguish one statement from the other:

If *expr* then (if *expr* then *stmt* else *stmt*)

If *expr* then (if *expr* then *stmt*) else *stmt*

It is impossible to determine which if condition the else applies.

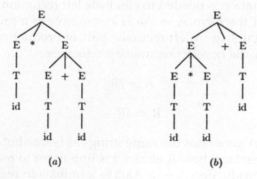

(a)   (b)

FIGURE 3.2
Parse trees for id * id + id.

There are multiple strategies for addressing this problem. One approach is to change the grammar so that it imposes the constraint that the else is attached to the nearest if

```
1.  stmt:= …
2.  | selection- stmt

3.  stmt-with-else:= …
4.  | selection-stmt-with-else

5.  selection- stmt:= …
6.  | IF (expr) stmt
7.  | IF (expr) stmt-with-else ELSE stmt

8.  selection- stmt-with-else:= …
9.  | IF (expr) stmt-with-else ELSE stmt-with-else
```

The production bodies in lines 6 and 7 force the else to go with the nearest if when coupled with the body in line 9.

Although compiler designers rarely do so for a complete programming language grammar, it is useful to be able to reason that a given set of productions generates a particular language. Troublesome constructs can be studied by writing a concise, abstract grammar and studying the language that it generates. A proof that a grammar G generates a language L has two parts:

1. Show that every string generated by G is in L.

2. Conversely show that every string in L can indeed be generated by G.

This is a very difficult problem.

3.4.4 Removing Left Recursion

As stated previously, a grammar is left recursive if it has a nonterminal A such that there is a derivation $A + \Rightarrow Aa|\beta$ for some string a. This is a problem because top-down parsing methods cannot handle left-recursive grammars. A transformation is needed to eliminate left recursion.

A production of the form $A \Rightarrow Aa$ is an example of a production that is immediate left recursive. A left-recursive pair of productions $A + \Rightarrow Aa|\beta$ can be replaced by the non-left-recursive productions:

$$A := \beta R$$

$$R := aR|\in$$

The nonterminal A generates the same strings as before but is no longer left recursive. To understand how it works, it is important to recognize that the only way for the production $A + \Rightarrow Aa|\beta$ to terminate in recursion is for the production A ⇒ β to be applied. Upon doing that, β will be the first item in the sequence and it will followed by sequence of a's or nothing.

Example 3.17

Consider the following grammar:

$$E := E + T | E - T | T$$

$$T := T * F | T / F | F$$

$$F := (E) | id$$

This grammar is left recursive in both E and T. Transforming it into right-recursive productions, the production E can be rewritten as

$$E := T\ E'$$

with E' given by the productions:

$$E' := + T\ E' | -T\ E'$$

This eliminates the left recursion over E, but not T. Transforming it into right-recursive productions, the production T can be rewritten as

$$T := F\ T'$$

$$T' := * F\ T' | / F\ T'$$

The final production, F := (E)|id, is not recursive so it remains as it was. The grammar is now

$$E := T\ E'$$

$$E' := + T\ E' | -T\ E' | \in$$

$$T := F\ T'$$

$$T' := * F\ T' | / F\ T' | \in$$

$$F := (E) | id$$

This procedure eliminated all left recursion from immediate left-recursive productions, but it does not eliminate left recursion involving derivations of two or more steps.

A subtler form of left recursion can creep into a grammar. For example, consider the following grammar:

$$S := A\,a\,|\,b$$

$$A := A\,c\,|\,S\,d\,|\,\in$$

The nonterminal S is left recursive because $S \Rightarrow Aa \Rightarrow Sda$. However, it is not immediately left recursive because it requires the intermediate derivation involving the production for A. A grammar could also have cycles. For example, the following grammar has the cycle $S \Rightarrow Y \Rightarrow X \Rightarrow S$:

$$S := Xb\,|\,Y$$

$$Y := X\,|\,a$$

$$X := S\,|\,c$$

To convert an indirect left recursion production rule into one with right recursion, a means of identifying all places where indirect recursion may occur is necessary. Once a place where indirect recursion occurs has been identified, then it is necessary to apply the right recursion transformation to remove that place. An algorithm to eliminate all left recursion from a grammar does so by thorough application of two techniques:

1. Forward substitution to convert indirect left recursion into direct left recursion.
2. Rewriting direct left recursion as right recursion.

This approach assumes the original grammar does not have cycles ($A \Rightarrow^* A$) or $\in$-productions.

Arrange the nonterminals in some arbitrary order, A1, A2, ..., An.

For i = 1 to n {
       For j = 1 to i–1 {
              Replace each production of the form Ai := Aj γ by the productions Ai := $\delta1\,\gamma\,|\,\delta2\,\gamma\,|...|\,\,\delta$k γ, where Aj := $\delta1\,|\,\delta2\,|...|\,\delta$k.
       }
       Eliminate any immediate left recursion among the newly introduced productions.
}

The algorithm first imposes an arbitrary order on the nonterminals. This is a bookkeeping step. The outer loop cycles through the nonterminals in the selected order. It uses the right recursion transformation to remove any immediate left-recursive productions. The inner loop looks for any production that expands Ai into a right-hand side that begins with Aj, for j < i. Such an expansion may lead to an indirect left recursion. To avoid this, the algorithm replaces the occurrence of Aj with all the alternative right-hand sides for Aj. That is, if the inner loop discovers a production Ai := Aj γ, where Aj =: δ1|δ2|...|δk, then the algorithm replaces Ai := Aj γ with a set of productions Ai := δ1 γ|δ2 γ|...|δk γ. This step exposes any indirect left-recursive productions by converting each indirect left-recursive production into a direct left-recursive production. The final step in the outer loop converts any direct left recursion on Ai to a right-recursive production using the simple transformation shown earlier. Because new nonterminals are added at the end and only involve right recursion, it is unnecessary to check and convert them. Thus, all the direct and indirect left-recursive productions are eliminated.

3.4.5 Left Factoring

Left factoring is a grammar transformation used to produce a grammar suitable for top-down parsing. It comes into play when the choice between two alternative A productions is not clear. In this case, it may be possible to rewrite the productions so that the decision is deferred until sufficient input has been processed to make the right choice.

Consider the following two productions:

$$A := a \, \beta_1$$

$$A := a \, \beta_2$$

When processing the input, it is not possible to immediately tell which production to choose to expand. In general, if $A + \Rightarrow a \, \beta_1 | a \, \beta_2$ and the input begins with a nonempty string derived from a, then it is not possible to establish whether to expand A to a β_1 or a β_2.

It may be possible to defer the decision by expanding A to aA' where A' is a production of the form $A' := \beta_1 | \beta_2$. Thus, after seeing the input derived from a, it is then possible to expand A' to a β_1 or to a β_2. That is, left-factored, the original productions become

$$A := a \, A'$$

$$A' := \beta_1 | \beta_2$$

3.5 Regular Expressions versus Grammars

While regular expressions and grammars both deal with specifying languages, it should be noted that grammars are a more powerful notation than regular expressions. Any construct that can be described by a regular expression can be described by a grammar. The opposite is not true. Every regular language (as defined by regular expressions) is a context-free language, but not every context-free language is a regular language. For example, the regular expression (a|b) * abb and the grammar

$$A := Aabb$$

$$A := a|b$$

describe the same language, the set of strings of a's and b's ending in abb.

On the other hand, the language L = $\{a^n b^n | n \geq 1\}$ with an equal number of a's and b's is a prototypical example of a language that can be described by a grammar but not by a regular expression. To see why, suppose L were the language defined by some regular expression. We could construct a DFA D with a finite number of states, say k, to accept L. Since D has only k states, for an input beginning with more than k a's, D must enter some state twice, say S_i. Suppose that the path from S_i back to itself is labeled with a sequence a^{i-j}. Since $a^i b^i$ is in the language, there must be a path labeled b^i from S_i to an accepting state f. But then there is also a path from the initial state S_0 through S_i to f labeled $a^j b^i$. Thus, D also accepts $a^j b^i$, which is not in the language, contradicting the assumption that L is the language accepted by D.

We say that "finite automata cannot count," meaning that a finite automaton cannot accept a language like L = $\{a^n b^n | n \geq 1\}$ that would require it to keep count of the number of a's before it sees the b's. Likewise, a grammar can count two items but not three. Grammars can describe most, but not all, of the syntax of programming languages. For instance, the requirement that identifiers be declared before they are used cannot be described by a context-free grammar. Therefore, the sequences of tokens accepted by a parser form a superset of the programming language; subsequent phases of the compiler must analyze the output of the parser to ensure compliance with rules that are not checked by the parser.

If everything that can be described by a regular expression can also be described by a grammar, then why use regular expressions to define the lexical syntax of a language? There are several reasons. Separating the syntactic structure of a language into lexical and nonlexical parts provides a convenient way of modularizing the front end of a compiler into two manageable-sized components. The lexical rules of a language are frequently quite simple, and to describe them we do not need a notation as powerful as grammars. Regular expressions generally provide a more concise and easier-to-understand notation for tokens than grammars. More efficient lexical analyzers can be constructed automatically from regular expressions than from arbitrary grammars.

A few syntactic constructs found in typical programming languages cannot be specified using grammars alone. For example, no grammar or regular expression can ensure that

- Variables are declared before they are used.
- The proper number of arguments appear in a function call.

That is a problem left to the designer of a compiler.

Exercises

1. Let $A = \{a, b, c\}$ and $B = \{\alpha, \beta, \gamma\}$. List $A \cup B$ and AB.
2. Let $A = \{1, 2\}$ and $B = \{0-9\}$. List $A \cup B$ and AB.
3. Let $s = abc$. Give s^0, s^1, s^2, and s^3.
4. The following rules describe a formal language L over the alphabet $\Sigma = \{0,1,2,3,4,5,6,7,8,9,+,=\}$:
   - Every nonempty string that does not contain "+" or "=" and does not start with "0" is in L.
   - The string 0 is in L.
   - A string containing "=" is in L if and only if there is exactly one "=", and it separates two valid strings of L.
   - A string containing "+" but not "=" is in L if and only if every "+" in the string separates two valid strings of L.
   - No string is in L other than those implied by the previous rules.

   Which of the following strings are in L and which are not?
   a. 0123
   b. 1230
   c. 89+
   d. +89
   e. 0 + 89
   f. 567=
   g. 567 = 0
   h. 23 + 1 = 12
   i. 23 + 1 = 24
   j. 23 − 1 = 22

5. Assume that the language L is the alphabet {0, 1}. Give its Positive Closure and Kleene Closure.

6. Assume that the language L is the alphabet {α, β, γ}. Give its Positive Closure and Kleene Closure.

7. Given regular expressions
   a. ab*
   b. (a|b)*
   c. a|(b|ε)
   d. a|(b|ε)*
   e. (a|ε)*|(b|ε)*
   f. a?[abc]
   g. [ab]?(cb)*

   Describe the language that each of them specifies.

8. Given regular expressions
   a. 0|1
   b. 0(0|1)1
   c. 0(0|1)*1
   d. 0(0|1|ε)*
   e. (0|1|ε)*(1|ε)*
   f. 1?[01]+
   g. [01]?[01]*[01]

   Describe the language that each of them specifies.

9. Let E be an expression, T a term and id an identifier. Given the grammar

$$E := E + T | E-T | -E | (E) | T$$

$$T := T * T | T/T | id$$

   Give derivations of the following sentences:
   a. E * ⇒ (−id) + id
   b. E * ⇒ −(id + id) + id* id
   c. E * ⇒ (id * id) + id/(id* id)
   d. E * ⇒ (id * id/id) + id/(id* id)

10. Consider the grammar on alphabet {0, 1}, with the start symbol S and the following two production rules:

$$S \rightarrow Sa$$

$$S \rightarrow b$$

Describe the language of the grammar.

11. Consider the grammar on alphabet {0, 1}, with the start symbol S and the following two production rules:

$$S \rightarrow 10S$$

$$S \rightarrow 1|0$$

Describe the language of the grammar.

12. Consider the grammar

$$S \rightarrow S + S|S - S|a$$

Show there are two leftmost derivations for the string a + a −a and draw their parse trees.

13. Consider the grammar

$$S \rightarrow A-A$$

$$A \rightarrow a|b$$

a. Explain the language of the grammar.

b. How many derivations are there for the string a–b? How many of them are leftmost?

14. A credit card number is 16 digits, and is typically divided into 4 groups of 4 digits, although it is also occasionally written as one big number. Follow Example 3.7 to develop a regular expression that can validate credit card numbers in any of the following forms:

0123456789012345

0123 4567 8901 2345

0123-4567-8901-2345

0123,4567,8901,2345

15. A valid American telephone number starts with a 3-digit area code, followed by a 3-digit central office prefix, and then followed by a 4-digit line number. Follow Example 3.7 to develop a regular expression that can validate telephone numbers in any of the following forms:

    0123456789

    012 345 6789

    012-345-6789

    (012)345-6789

4

Propositional Logic

Logic plays a fundamental role in computer science. It takes into account syntactically well-formed statements and studies whether they are semantically correct. Propositional logic deals with declarative sentences, or propositions. Propositions can be formed by other propositions with the use of logical operators. Propositional logic is concerned with the study of the truth value of propositions and how their value depends on the truth or falsity of their component propositions.

4.1 Propositional Statements

A *proposition* is a declarative sentence, or part of a sentence, that can be evaluated to *true* or *false*, but not both. For example, the following sentences are propositions:

- *Earth is the center of the universe.*
- *Five plus five is equal to ten and five minus five is equal to zero.*
- *If it is raining, then the ground is wet.*

Here the first sentence is false according to our knowledge on the universe. The second sentence is true according to mathematics. The third sentence is also true. The following sentences, however, are not propositions:

- *Is it raining?*
- *Let us go!*
- *Please take an umbrella with you.*

They are not declarative sentences and are not assertion on any facts.

An indivisible proposition is called an *atom*, or *atomic proposition*. For example, the proposition "Earth is the center of the universe" is indivisible, and thus it is an atom. The proposition "If it rains, then the ground is wet" is composed of two atoms:

- *It is raining.*
- *The ground is wet.*

These two atoms are composed into one proposition with the "if ... then ..." connective.

Example 4.1: Atomic Propositions

List all atoms of the following proposition:

If it snows or the temperature is over 100°F, then the school is closed.

The solution is that there are three atoms in the proposition:

- *It snows.*
- *The temperature is over 100°F.*
- *The school is closed.*

An atom cannot be broken down; otherwise there will be a change to its meaning. For example, the sentence *Joe and John are brothers* is an atom. If we break it into "Joe is a brother" and "John is a brother," the meaning in the original sentence is changed. The sentence *Joe and John study hard*, however, is divisible and can be broken into "Joe studies hard" and "John studies hard."

Propositional logic focuses on the composition of logic sentences and truth value of the resulting compounds. It does not study the context of logic sentences. As such, atomic propositions are denoted by letters to avoid vagueness and equivocation of nature languages.

Example 4.2: Use Symbols for Atoms

Consider the following atomic sentences:

p: It snows.
q: The temperature is over 100°F.
r: The school is closed.

The compound proposition in Example 4.1 can be rewritten as

If p or q, then r.

4.2 Logic Operators and Truth Tables

As shown in Example 4.1, atomic propositions can be combined to form compound propositions. In logic, a *logic operator* (also called *logic connective*) is a symbol or word used to connect one or more sentences in a way that the value of the compound sentence produced depends only on that of the original sentences and on the meaning of the connective. There are four propositional logical operators:

1. ¬ denotes negation, or *"not"*
2. ∧ denotes conjunction, or *"and"*

3. ∨ denotes disjunction, or *"or"*

4. → denotes implication, or *"if-then"*

Example 4.3: Logic Operators

Given the following atoms:

p: *Joe studies hard.*
q: *John studies hard.*

The following are compound sentences when logic operators are applied to these atoms:

¬*p*: *Joe does not study hard.*
p ∧ *q*: *Joe studies hard and John studies hard.*
p ∨ *q*: *Joe studies hard or John studies hard.*
p ∨ ¬*p*: *Either Joe studies hard or he does not study hard.*
p → *q*: *If Joe studies hard, then John studies hard.*
¬*p* → *q*: *If Joe does not study hard, then John studies hard.*

With the four propositional operators, we can construct a logic formula at any level of complexity from atoms. Among the operators, ¬ is unary and has the highest binding priority. Both ∧ and ∨ are binary operators and bind more tightly than →. The implication operator → has the lowest binding priority and is right associative. When evaluating a formula, we should follow the binding priorities of logic operators to proceed step-by-step.

Example 4.4: Binding Priorities of Propositional Operators

The formula $p \wedge \neg q \vee r \to s$ is equivalent to

$$((p \wedge (\neg q)) \vee r) \to s$$

Because → is right associative, the formula $p \wedge \neg q \to r \to s \to t$ is equivalent to

$$((p \wedge (\neg q)) \to (r \to (s \to t))$$

4.2.1 Well-Formed Formulas

We call compound propositions *propositional formulas*, or simply formulas. As a convention, we use lowercase English letters for atomic propositions and the Greek alphabet for propositional formulas. A formula is *well-formed* if it is composed using the following inductive construction rules:

- An atomic proposition p is a well-formed formula.
- If φ is a well-formed formula, then so is $\neg\varphi$.

- If φ and ψ are well-formed formulas, then so are $\varphi \land \psi$, $\varphi \lor \psi$, and $\varphi \rightarrow \psi$.
- If φ is a well-formed formula, then so is (φ).

This definition can also be described using BNF (Backus-Naur form) notation as follows:

```
<formula>::= <atomic proposition>
           | ¬<formula>
           | <formula> ∧ <formula>
           | <formula> ∨ <formula>
           | <formula> → <formula>
           | (<formula>)
```

or simply

$$\varphi ::= p \mid (\neg \varphi) \mid (\varphi \land \varphi) \mid (\varphi \lor \varphi) \mid (\varphi \rightarrow \varphi) \qquad (4.1)$$

Example 4.5: Well-Formed Formulas

The following propositional formulas are all well-formed:

1. $p \land \neg q \lor r \rightarrow s$
2. $\neg p \rightarrow q \lor r \rightarrow s$
3. $\neg p \land r \rightarrow q \lor \neg r \land s$
4. $\neg p \land r \rightarrow ((q \rightarrow \neg r) \land s)$
5. $\neg (p \lor r) \rightarrow (q \lor \neg r \rightarrow s) \land (q \rightarrow \neg r)$

They can all be constructed recursively with the rules in formula 4.1. For example, we can construct formula 5 sequentially with the following steps:

$p \lor r$

$\neg (p \lor r)$

$\neg r$

$q \lor \neg r$

$q \lor \neg r \rightarrow s$

$q \rightarrow \neg r$

$(q \lor \neg r \rightarrow s) \land (q \rightarrow \neg r)$

$\neg (p \lor r) \rightarrow (q \lor \neg r \rightarrow s) \land (q \rightarrow \neg r)$

The following formulas are not well-formed:

1. $p \wedge \neg q \vee \rightarrow s$
2. $\neg p \rightarrow q \neg r \rightarrow s$
3. $\wedge r \rightarrow q \vee \neg r \wedge s$

In formula 1, the disjunction operator has only one valid operand. In formula 2, the second negation operator is associated with two operands. In formula 3, the first conjunction operator does not have a left operand.

The well-formedness of a formula can also be checked through its parse tree. A parse tree of a formula is built up by applying the inductive formula construction rules in reverse. It starts by breaking down the formula through the main operator, which is the operator that is applied at the last step in the formula construction, or equivalently the operator that is to be processed at the last step during the formula evaluation. For example, consider

$$\varphi: \neg(p \vee r) \rightarrow (q \vee \neg r \rightarrow s) \wedge (q \rightarrow \neg r)$$

The main operator in φ is the $\rightarrow$ that connects the following two subformulas:

$$\neg(p \vee r)$$

$$(q \vee \neg r \rightarrow s) \wedge (q \rightarrow \neg r)$$

Then for each subformula, we further break it down through the main operator. The main operator in the first subformula is $\neg$, while the one in the second subformula is $\wedge$. Therefore, at the next level of the parse tree, we get the following further subformulas:

$$p \vee r$$

$$q \vee \neg r \rightarrow s$$

$$q \rightarrow \neg r$$

We repeat the process until we arrive at only propositional atoms. Figure 4.1 shows the parse tree of φ. In case there is no confusion, we can simply use operators to be evaluated for all non-leaf nodes.

A parse tree represents a well-formed formula if

1. The root is an atom and nothing else (i.e., the formula is a p), or
2. The root is $\neg$ and there is only one well-formed subtree, or
3. The root is $\wedge$, $\vee$, or $\rightarrow$ and there are two well-formed subtrees.

Notice in the parse tree of a well-formed formula, all leaf nodes are atoms.

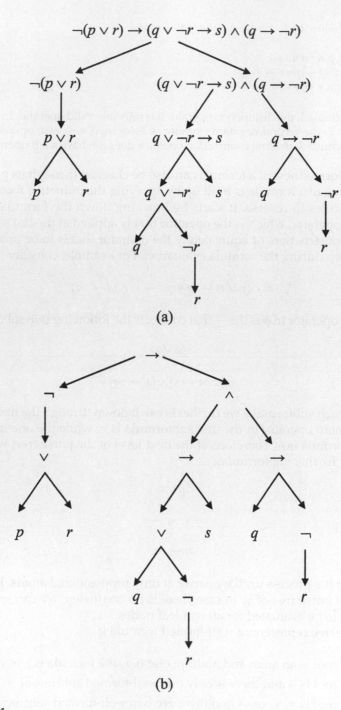

FIGURE 4.1

(a) Parse tree of $\neg(p \vee r) \rightarrow (q \vee \neg r \rightarrow s) \wedge (q \rightarrow \neg r)$. (b) A simplified version.

4.2.2 Truth Values

The semantics of each propositional logic operator is defined through a truth table. Let p be a propositional statement. The negation of p, denoted by $\neg p$, states what it would mean for p to be false. Therefore, $\neg p$ has opposite truth value from p: if p is true, then $\neg p$ is false; if p is false, then $\neg p$ is true. The truth table for negation is shown in Table 4.1, in which T represents true and F represents false.

The conjunction (*and*) of two propositional statements p and q is denoted by $p \wedge q$. $p \wedge q$ is true when both p and q are true. As long as one of the two component statements is false, then their conjunction is false. The truth values of conjunction are listed in Table 4.2.

The disjunction (*or*) of two propositional statements p and q is denoted by $p \vee q$. $p \vee q$ is true when p or q or both are true. The disjunction is false only when both p and q are false. The truth values of disjunction are listed in Table 4.3.

TABLE 4.1

Truth Table for Negation

p	$\neg p$
T	F
F	T

TABLE 4.2

Truth Table for Conjunction

p	q	$p \wedge q$
T	T	T
T	F	F
F	T	F
F	F	F

TABLE 4.3

Truth Table for Disjunction

p	q	$p \vee q$
T	T	T
T	F	T
F	T	T
F	F	F

The implication operator "→" constructs conditional statements. In the symbolic representation $p \rightarrow q$, p represents *hypothesis* and q stands for *conclusion*. $p \rightarrow q$ is false only in the case that p is true and q is false. The reason behind this is that a conditional statement is a simple *reasoning* statement: if the hypothesis part is true, the conclusion should also be true. The reasoning fails if the hypothesis is true but the conclusion is false. However, if the hypothesis is false, the value of the conclusion is no longer important because the reasoning here is only about conveying the truth of the hypothesis to the conclusion. The reasoning is never violated when the hypothesis is not true, for there is no truth to be conveyed to the conclusion. Consider the following proposition:

> *If you fall within the top 20% of your major program by semester GPA, then you will be recognized on the dean's list for that semester.*

The proposition is false if you fall within the top 20% of your major program by semester GPA, but you are not recognized on the dean's list for that semester. However, if your semester GPA is not within the top 20%, you cannot say the statement is not true, regardless of whether you are or you are not recognized on the dean's list.

The truth values of implication are listed in Table 4.4.

4.2.3 Evaluation of Propositional Formulas

In evaluating a propositional formula, each atomic proposition in the formula is treated as a variable. Evaluation begins with assignment of a truth value to each variable. Each truth value assignment is called an *interpretation* or *model*. Because each variable represents a simple sentence, the truth values are being applied to the "truth" or "falsity" of these simple sentences. For example, each row of Table 4.4 corresponds to an assignment. Given a formula, we need to evaluate it for all possible assignments. If the formula has n variables, then the truth table that evaluates the formula will have 2^n rows.

TABLE 4.4

Truth Table for Implication

p	q	$p \rightarrow q$
T	T	T
T	F	F
F	T	T
F	F	T

Example 4.6: Evaluation of Formulas

Construct the truth tables of the following formulas:

$$\varphi_1 = \neg p \vee q$$

$$\varphi_2 = \neg p \rightarrow \neg q \rightarrow p \vee q$$

$$\varphi_3 = p \wedge \neg q \vee r \rightarrow (p \vee \neg r)$$

Solution

To construct the truth table of a formula, we take all component atomic propositions of the formula as variables, and then evaluate the formula by assigning all different combinations of truth values to the variables. For example, the formula φ_1 has two component variables: p and q, and thus we evaluate φ_1 for each of the four possible truth value assignments of p and q. In case a formula is complex, we can break it into several sub-formulas and first evaluate each subformula, and then evaluate the entire formula. For example, for φ_3, we can evaluate $p \wedge \neg q \vee r$ and $(p \vee \neg r)$ first. By the same token, we can evaluate $p \wedge \neg q$ before we evaluate $p \wedge \neg q \vee r$.

Tables 4.5 through 4.7 show the truth tables for φ_1, φ_2, and φ_3. respectively.

Comparing Table 4.5 with Table 4.4, we can tell that $p \rightarrow q$ has the same truth values as $\neg p \vee q$ for any truth value assignment to p and q. Yes, $p \rightarrow q$ is equivalent to $\neg p \vee q$, and it is denoted by

$$p \rightarrow q \equiv \neg p \vee q$$

TABLE 4.5

Truth Table for $\varphi_1 = \neg p \vee q$

p	q	$\neg p$	φ_1
T	T	F	T
T	F	F	F
F	T	T	T
F	F	T	T

TABLE 4.6

Truth Table for $\varphi_2 = \neg p \rightarrow \neg q \rightarrow p \vee q$

p	q	$\neg p$	$\neg q$	$p \vee q$	$\neg q \rightarrow p \vee q$	φ_2
T	T	F	F	T	T	T
T	F	F	T	T	T	T
F	T	T	F	T	T	T
F	F	T	T	F	F	F

TABLE 4.7

Truth Table for $\varphi_3 = p \wedge \neg q \vee r \rightarrow (p \vee \neg r)$

p	q	r	$\neg q$	$\neg r$	$p \wedge \neg q$	$p \wedge \neg q \vee r$	$p \vee \neg r$	φ_3
T	T	T	F	F	F	T	T	T
T	T	F	F	T	F	F	T	T
T	F	T	T	F	T	T	T	T
T	F	F	T	T	T	T	T	T
F	T	T	F	F	F	T	F	F
F	T	F	F	T	F	F	T	T
F	F	T	T	F	F	T	F	F
F	F	F	T	T	F	F	T	T

This is an important property, as sometimes we need to convert one form of the proposition to another form.

There are two special types of logic formulas. One is in the form $\varphi \vee \neg\varphi$ and always evaluates to true. We call it a *tautology* and denote it by T (reads "top"). The other one is in the form $\varphi \wedge \neg\varphi$ and always evaluates to false. We call it a *contradiction* and denote it by $\bot$ (reads "bottom").

4.3 Logic Equivalencies

Two statements are logically equivalent if they have the same truth value in every combination of truth value assignments to component atoms. We have shown that $p \rightarrow q \equiv \neg p \vee q$ in the previous section. We introduce more equivalencies in this section. Many of the equivalences are known as *laws*.

Identity laws:

$$p \wedge T \equiv p$$

$$p \vee F \equiv p$$

Domination laws:

$$p \vee T \equiv T$$

$$p \wedge F \equiv F$$

Idempotent laws:

$$p \vee p \equiv p$$

$$p \wedge p \equiv p$$

Double negation law:

$$\neg\neg p \equiv p$$

Commutative laws:

$$p \wedge q \equiv q \wedge p$$

$$p \vee q \equiv q \vee p$$

Associative laws:

$$p \wedge (q \wedge r) \equiv (p \wedge q) \wedge r$$

$$p \vee (q \vee r) \equiv (p \vee q) \vee r$$

Distributive laws:

$$p \wedge (q \vee r) \equiv (p \wedge q) \vee (p \wedge r)$$

$$p \vee (q \wedge r) \equiv (p \vee q) \wedge (p \vee r)$$

De Morgan's laws:

$$\neg(p \wedge q) \equiv \neg p \vee \neg q$$

$$\neg(p \vee q) \equiv \neg p \wedge \neg q$$

Absorption laws:

$$p \wedge (p \vee q) \equiv p$$

$$p \vee (p \wedge q) \equiv p$$

Law of excluded middle:

$$p \vee \neg p \equiv \text{T}$$

Contrapositive of implication:

$$p \to q \equiv \neg q \to \neg p$$

Negation of implication:

$$\neg(p \to q) \equiv p \wedge \neg q$$

Example 4.7: De Morgan's Law

Show De Morgan's law:

$$\neg(p \wedge q) \equiv \neg p \vee \neg q$$

Solution

We show the law with truth table. As listed in Table 4.8, $\neg(p \wedge q)$ and $\neg p \vee \neg q$ have the same truth value in every truth value assignment to p and q. Therefore, these two formulas are equivalent.

Example 4.8: Contrapositive of Implication

Show the law of contrapositive of implication

$$p \to q \equiv \neg q \to \neg p$$

Solution

$$p \to q \equiv \neg p \vee q$$

$$\equiv q \vee \neg p \text{ (Commutative law)}$$

$$\equiv \neg\neg q \vee \neg p \text{ (Double negation law)}$$

$$\equiv \neg(\neg q) \vee (\neg p)$$

$$\equiv \neg q \to \neg p$$

TABLE 4.8

Truth Table for $\neg(p \wedge q)$ and $\neg p \vee \neg q$

p	q	$\neg p$	$\neg q$	$p \wedge q$	$\neg(p \wedge q)$	$\neg p \vee \neg q$
T	T	F	F	T	F	F
T	F	F	T	F	T	T
F	T	T	F	F	T	T
F	F	T	T	F	T	T

Consider the following statement:

If it is raining, then the ground is wet.

It is equal to say

If the ground is not wet, then it is not raining.

Example 4.9: Negation of Implication

Show the law of the negation of implication:

$$\neg(p \to q) \equiv p \wedge \neg q$$

Solution

$$\neg(p \to q) \equiv \neg(\neg p \vee q)$$

$$\equiv \neg\neg p \wedge \neg q \text{ (De Morgan's law)}$$

$$\equiv p \wedge \neg q \text{ (Double negation law)}$$

4.4 Logic Arguments

A logic argument is a sequence of propositions. The last proposition in the sequence is called a *conclusion*, while all other propositions are called *premises* or *assumptions*. An argument is stated in a *logic form*, which is the logic structure of the sequence of propositions. One can represent the logical form of an argument by replacing the specific content words with letters used as placeholders or variables. For example, the argument

If turtles are reptiles, then turtles are cold-blooded.
Turtles are reptiles.
Therefore, turtles are cold-blooded.

has the following logic form:

If p then q
p
Therefore, q

We use the symbol ⊢ to represent "therefore," the connector between the premises and conclusion. The above logic form can be further abstracted as

$$p \to q, p \vdash q$$

4.4.1 Validity and Soundness

An argument is *valid* if when all the premises are true, the conclusion is also true. In a valid argument, the premises and conclusion are related to each other in the right way so that if the premises *were* true, then the conclusion would have to be true as well. Simply put, the premises provide support for the conclusion in a valid argument.

It should be stressed that whether or not an argument is valid is purely determined based on its logic form. It is irrelevant to the actual truth or falsity of its premises and conclusion. In other words, a valid argument could have a false conclusion, and an invalid argument could have a true conclusion.

> **Example 4.10: Validity of arguments**
>
> Consider the following argument:
>
> > *If all dogs can fly, then all dogs are birds.*
> > *All dogs can fly.*
> > *Therefore, all dogs are birds.*
>
> In this argument, the falsity of the second premise leads to the falsity of the conclusion. However, the argument is valid, because the argument form is valid: when you assume the two premises are true, then the conclusion is also true. Here is a *good* argument in the same logic form:
>
> > *If all dogs can walk, then all dogs are animals.*
> > *All dogs can walk.*
> > *Therefore, all dogs are animals.*
>
> The following argument is invalid:
>
> > *If all dogs can walk, then all dogs are animals.*
> > *All dogs are animals.*
> > *Therefore, all dogs can walk.*
>
> This argument has the following logic form:
>
> > *If p then q*
> > *q*
> > *Therefore, p*
>
> We know the conclusion of the argument is true; however, the argument is invalid because its logic form is invalid. Take a look at Table 4.4 for implication. The third row indicates when $p \rightarrow q$ and q are true, p is false, a case of violation of the argument.

An argument is *sound* if it is valid and meanwhile all of its premises are true. For example, the first argument in Example 4.10 is valid but not sound. The second "good" argument is both valid and sound.

4.4.2 Validity Test through Truth Tables

The validity of an argument form can be tested by constructing the truth table for the argument form. The process starts with the identification of all propositional variables in the form. The next step is to construct a truth table that shows the truth values of all the premises and the conclusion for all possible truth value assignments of the propositional variables. If there is a row in the truth table in which all premises are true but the conclusion is false, then it is possible for an argument of the form to have true premises and false conclusion, and thus the argument form is invalid. If such a row does not exist in the table, then it means whenever all premises are true the conclusion is also true, and thus the argument form is valid.

Example 4.11: A Valid Argument Form

Table 4.9 shows that the argument form $\neg p \rightarrow q, \neg q \vdash p$ is valid. There is only one row (highlighted) in which the two premises are true, and the conclusion is also true in the row.

Example 4.12: An Invalid Argument Form

Table 4.10 shows that the argument form $\neg p \rightarrow q, p \vdash \neg p \vee q$ is invalid. There are two rows (highlighted) in which the two premises are true. However, in the row where p is true and q is false, the premises are true but the conclusion is false. Therefore, the argument is invalid.

TABLE 4.9

Truth Table for $\neg p \rightarrow q, \neg q \vdash p$

p	q	$\neg p$	$\neg p \rightarrow q$	$\neg q$
T	T	F	T	F
T	F	F	T	T
F	T	T	T	F
F	F	T	F	T

TABLE 4.10

Truth Table for $\neg p \rightarrow q, p \vdash \neg p \vee q$

p	q	$\neg p$	$\neg p \rightarrow q$	$\neg p \vee q$
T	T	F	T	T
T	F	F	T	F
F	T	T	T	T
F	F	T	F	T

4.4.3 Inference Rules

Using truth tables to test the validity of an argument form is quite straightforward. The only inconvenience is that the number of truth value combinations of variables in a form grows exponentially with the number of variables. A valid argument can also be proved by applying some well-established inference rules and/or logic equivalences. We will describe all inference rules in the following format:

$$\frac{\text{premise 1, premise 2, ..., premise}\, m}{\text{conclusion}} \text{ abbrev. of rule name}$$

4.4.3.1 Or-Introduction (∨i)

This is a generalization rule. It follows the fact that if a statement is true, then the disjunction of this statement and a second statement will also be true, regardless of whether the second statement is true or false. The rule is represented as

$$\frac{\varphi}{\varphi \vee \psi} \vee i$$

4.4.3.2 And-Elimination (∧e)

This is a specialization rule. It follows the fact that if the conjunction of two statements is true, then each of the statements must be true. The rule is represented as

$$\frac{\varphi \wedge \psi}{\varphi} \wedge e \qquad \frac{\varphi \wedge \psi}{\psi} \wedge e$$

4.4.3.3 And-Introduction (∧i)

If two statements are true, then the conjunction of them is also true. The rule is represented as

$$\frac{\varphi, \psi}{\varphi \wedge \psi} \wedge i$$

4.4.3.4 Modus Ponens (MP)

Modus ponens, Latin for "the way that affirms by affirming" and abbreviated as MP, is the inference rule that if a conditional statement $p \rightarrow q$ is accepted, and the condition p holds, then the conclusion q is inferred. The first two arguments in Example 4.10 are of the form of MP. Notation-wise, the rule is represented as

$$\frac{\varphi \to \psi, \varphi}{\psi} \text{MP}$$

Because MP eliminates the implication operator, the rule is also called the *implication elimination*.

Example 4.13: Proof with MP

Show

$$p \to q, p \land \neg r \vdash q \land \neg r$$

Proof:
The proof starts by listing all premises, and then proceeds by working toward the conclusion. Each step is listed on a separate line, with the inference rule applied annotated.

1. $p \to q$	Premise
2. $p \land \neg r$	Premise
3. p	$\land e, 2$
4. $\neg r$	$\land e, 2$
5. q	MP, 1, 3
6. $q \land \neg r$	$\land i, 5, 4$

4.4.3.5 Modus Tollens (MT)

Modus tollens, Latin for "the way that denies by denying" and abbreviated as MT, is the inference rule that if a conditional statement $\varphi \to \psi$ is accepted, and the conclusion ψ is denied ($\neg \psi$ is true), then the condition φ is also denied ($\neg \varphi$ is true). This is essentially the law of contrapositive of negation. Notation-wise, the rule is represented as

$$\frac{\varphi \to \psi, \neg \psi}{\neg \varphi} \text{MT}$$

Example 4.14: Proof with MT

Show

$$p \to \neg r, q \to r, p \vdash p \land \neg q.$$

Proof:
1. $p \to \neg r$	Premise
2. $q \to r$	Premise

3. p Premise
4. $\neg r$ MP, 1, 3
5. $\neg q$ MT, 2, 4
6. $p \wedge \neg q$ $\wedge i$, 3, 5

4.4.3.6 Implication-Introduction (→i)

This rule states that if after assuming φ is true one can prove ψ is true, then $\varphi \rightarrow \psi$ is inferred. The rule is specified as

$$\frac{}{\varphi \rightarrow \psi}\!\rightarrow\!i$$

The box is called an *assumption box*. Notice that in order to prove ψ, we can use other statements, which can be premises or intermediate results deduced from premises. We need to open an assumption box when we make an assumption on a statement. When the assumption box is closed, we get an implication from the assumed statement to the last statement in the box.

Example 4.15: Proof with Implication-Introduction

Show

$$p \rightarrow q, r \rightarrow s \vdash p \wedge r \rightarrow q \wedge s$$

Proof:
The conclusion of this argument is a conditional statement. After listing the two premises, we should make an assumption about the hypothesis of the conditional statement, and then work toward its conclusion. An assumption box is opened when we make the assumption, and it is closed when the conclusion is proved. The step-by-step proof process is listed as follows:

1. $p \rightarrow q$ Premise
2. $r \rightarrow s$ Premise
3. $p \wedge r$ Assumption
4. p $\wedge e$, 3
5. r $\wedge e$, 3
6. q MP, 1, 4
7. s MP, 2, 5
8. $q \wedge s$ $\wedge i$, 6, 7
9. $p \wedge r \rightarrow q \wedge s$ $\rightarrow i$, 3–8

4.4.3.7 Or-Elimination (∨e)

There are three or-elimination rules:

$$\frac{\varphi \vee \psi, \neg\varphi}{\psi} \vee e$$

$$\frac{\varphi \vee \psi, \neg\psi}{\varphi} \vee e$$

$$\frac{\varphi \vee \psi, \varphi \to \lambda, \psi \to \lambda}{\lambda} \vee e$$

The validity of the first two rules is obvious: that a disjunction is true means at least one of its operands is true. If we know one operand is false, then the other one must be true. For example,

Given an integer x, x is either even or odd.

x is not even.

Therefore, x is odd.

An example of the third rule is as follows:

Given an integer x, x is either even or odd.

If x is even, then there is an integer y such that x + y = 3.

If x is odd, then there is an integer y such that x + y = 3.

Therefore, given an integer x, there is an integer y such that x + y = 3.

The rule shows that if a disjunction of two propositions is true and you want to prove a third proposition, then you only need to prove that each of the operand proposition implies the third proposition. Because there are two implications to prove in the process, we will need to open two assumption boxes.

Example 4.16: Proof with Or-Elimination

Show

$$p \vee q, \neg q \vee r \vdash p \vee r$$

Proof:
The two premises are both disjunctions. We follow the third *or-elimination* rule to prove the argument. We can choose either of the two premises to apply the rule.

1.	$p \lor q$	Premise
2.	$\neg q \lor r$	Premise
3.	p	Assumption
4.	$p \lor r$	$\lor i, 3$
5.	$p \to p \lor r$	$\to i, 3\text{-}4$
6.	q	Assumption
7.	$\neg\neg q$	$\neg\neg, 6$
8.	r	$\lor e, 2, 7$
9.	$p \lor r$	$\lor i, 8$
10.	$q \to p \lor r$	$\to i, 6\text{-}9$
11.	$p \lor r$	$\lor e, 1, 5, 10$

4.4.3.8 Proof by Contradiction (PBC)

A logical *contradiction* is the conjunction of a proposition p and its negation $\neg p$. It is a fundamental law that a proposition and its negation cannot
both be true at the same time. Here are some examples of contradiction:

1. *Today is Monday and today is not Monday.*
2. *Nobody goes to that restaurant because it is always too crowded.*
3. *All even numbers can be divided by two, but some even numbers cannot be
   divided by two.*

Proposition 1 is a contradiction stated explicitly, while the propositions 2 and
3 are contradictions stated implicitly.

Proof by contradiction (PBC) is a form of indirect proof of a proposition.
It starts by assuming that the opposite proposition is true, and then shows
such an assumption leads to a contradiction. We use the symbol $\perp$ to represent a contradiction. The inference rule of PBC is formulated as

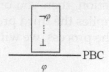

$$\frac{}{\varphi} \text{PBC}$$

Example 4.17: Proof by Contradiction

Show

$$\neg p \lor \neg q \vdash \neg(p \land q)$$

Proof:
The conclusion of the argument is $\neg(p \land q)$. To prove it with PBC, we
assume $p \land q$, and then try to arrive at a contradiction. The only premise
is a disjunction, so it is nature to apply the *or-elimination* rule to work
towards a contradiction.

1.	$\neg p \vee \neg q$	Premise
2.	$p \wedge q$	Assumption
3.	$\neg p$	Assumption
4.	p	$\wedge e, 2$
5.	$\bot$	Contradiction, 3, 4
6.	$\neg p \rightarrow \bot$	$\rightarrow i, 3\text{-}6$
7.	$\neg q$	Assumption
8.	q	$\wedge e, 2$
9.	$\bot$	Contradiction, 7, 8
10.	$\neg q \rightarrow \bot$	$\rightarrow i, 3\text{-}4$
11.	$\bot$	$\vee e, 1, 6, 10$
12.	$\neg(p \wedge q)$	PBC, 2–11

4.4.3.9 Law of the Excluded Middle (LEM)

This law was mentioned in Section 4.3 as a logic equivalence. It states that for any proposition, either it is true, or its negation is true. Therefore, we always have

$$\vdash \varphi \vee \neg\varphi$$

For example, if φ is the proposition

Socrates is mortal

then the LEM holds:

Socrates is mortal or Socrates is not mortal.

LEM is useful in proving arguments. It can be used as a premise. As using the *or-elimination* rule, we open an assumption box for the proposition and another one for its negation.

Example 4.18: Proof with LEM

Show

$$\vdash (p \rightarrow q) \vee (q \rightarrow r)$$

Proof:
This argument form has no premises. The conclusion is a disjunction of two operands. Based on the generalization rule, if we can prove one operand, then the disjunction is asserted. Because q appears in both operands, we use the LEM of q.

1.	$q \vee \neg q$	LEM
2.	q	Assumption
3.	$\neg p \vee q$	$\vee i, 2$
4.	$p \rightarrow q$	Equivalence, 3
5.	$(p \rightarrow q) \vee (q \rightarrow r)$	$\vee i, 4$
6.	$q \rightarrow (p \rightarrow q) \vee (q \rightarrow r)$	$\rightarrow i, 2\text{--}5$
7.	$\neg q$	Assumption
8.	$\neg q \vee r$	$\vee i, 7$
9.	$q \rightarrow r$	Equivalence, 8
10.	$(p \rightarrow q) \vee (q \rightarrow r)$	$\vee i, 9$
11.	$\neg q \rightarrow (p \rightarrow q) \vee (q \rightarrow r)$	$\rightarrow i, 7\text{--}10$
12.	$(p \rightarrow q) \vee (q \rightarrow r)$	$\vee e, 1, 6, 11$

4.5 Satisfiability of Formulas

A formula is *satisfiable* if there is a truth value assignment in which the formula evaluates to true. For example, the formula

$$(p \rightarrow q) \vee (q \rightarrow r)$$

is satisfiable, because it is true if we assign T's to all p, q and r. However,

$$(p \rightarrow q) \wedge p \wedge \neg q$$

is not satisfiable, because it is false in every truth value assignment. In fact, if we convert $p \rightarrow q$ to its equivalent $\neg p \vee q$, then the formula becomes

$$\neg p \vee q \wedge p \wedge \neg q$$

$$= (\neg p \vee q) \wedge p \wedge \neg q$$

$$= p \wedge (\neg p \vee q) \wedge \neg q$$

$$= (p \wedge \neg p) \vee (p \wedge q) \wedge \neg q$$

$$= (p \wedge q) \wedge \neg q$$

$$= \bot$$

So the formula is actually a contradiction, which is not satisfiable.

A formula is *valid* if for every truth value assignment it evaluates to true. For example, the formula

$$p \vee \neg p$$

is always true, and thus it is valid.

For a given formula φ, satisfiability and validity are dual properties:

- If φ is valid, then φ is satisfiable. However, if φ is satisfiable, then φ may or may not be valid.

- If $\neg\varphi$ is not valid, then φ is satisfiable. If $\neg\varphi$ is not satisfiable, then φ is valid.

Therefore, in order to find out if φ is satisfiable, we can check if $\neg\varphi$ is not valid. In order to find out if φ is valid, we can check if $\neg\varphi$ is not satisfiable.

From the formula construction rules one can construct a formula of arbitrarily complicated structure, and thus the validity or satisfiability check of the formula could be difficult. A *normal form* is a kind of simple form of formulas such that formulas in the form are easier to evaluate. One of the most common normal forms is conjunctive normal form. We will introduce it in the rest of the section and discuss how it simplifies the validity and satisfiability check of formulas.

4.5.1 Conjunctive Normal Forms

Any atomic proposition p or its negation $\neg p$ is called a *literal*. A formula is in *conjunctive normal form* (CNF) if it is a conjunction of clauses, where a clause is a disjunction of literals. For example, let

$$\varphi = (p \vee q \vee \neg r) \wedge (\neg p \vee r) \wedge s$$

$$\psi = (p \vee q \vee \neg r) \wedge (\neg p \vee r) \wedge (s \wedge t \vee r)$$

Then φ is in CNF because it is a conjunction of three clauses: $(p \vee q \vee \neg r)$, $(\neg p \vee r)$, and s, and each of the clauses is a disjunction of literals (s can be viewed as $s \vee s$). However, ψ is not in CNF, because the third clause $s \wedge t \vee r$ is not a disjunction of literals.

A formula is *valid* if it evaluates to true for every truth value assignment. Let φ be a formula and it is represented in CNF as

$$\varphi = \varphi_1 \wedge \varphi_2 \wedge \ldots \wedge \varphi_m$$

φ is valid if and only if every φ_i is valid, $i = 1, 2, \ldots, m$. The proof of this statement is quite straightforward: obviously, if every φ_i is valid, then based on the definition of conjunction, φ is valid. On the other hand, if φ is valid, then every φ_i must be valid, because if one clause is not valid, then there must be a truth assignment such that the clause evaluates to false, and thus φ evaluates to false.

Let

$$\varphi_i = l_1 \vee l_2 \vee \ldots \vee l_n$$

where l_j is either an atom or the negation of an atom, $j = 1, 2, \ldots, n$. φ_i is valid if and only if there are l_j and l_k such that $l_j = \neg l_k$. On the one hand, if $l_j = \neg l_k$, then because $l_j \vee \neg l_k$ is always true, so φ_i is always true. On the other hand, if such l_j and l_k do not exist, then under the assignment all literals are false, φ_i is false.

Any well-formed propositional formula can be converted to an equivalent formula in CNF. In general, the conversion takes three steps:

1. Eliminate $\rightarrow$ using the equivalency

$$p \rightarrow q \equiv \neg p \vee q$$

2. Move $\neg$ inward using the De Morgan's laws

$$\neg(p \vee q) \equiv \neg p \wedge \neg q$$

$$\neg(p \wedge q) \equiv \neg p \vee \neg q$$

3. Distribute $\vee$ inward over $\wedge$ using the distributive law

$$p \vee (q \wedge r) \equiv (p \vee q) \wedge (p \vee r)$$

Example 4.19: Formula in CNF

Convert the formula

$$\varphi = p \wedge \neg q \vee \neg r \rightarrow q \wedge \neg p \vee r \rightarrow s$$

to an equivalent formula in CNF.

First, we eliminate the two implication operators based on the equivalency $p \rightarrow q \equiv \neg p \vee q$. Because implication is right associative, we eliminate the right one first.

$$\varphi = (p \wedge \neg q \vee \neg r) \rightarrow ((q \wedge \neg p \vee r) \rightarrow s)$$

$$= (p \wedge \neg q \vee \neg r) \rightarrow (\neg(q \wedge \neg p \vee r) \vee s)$$

$$= \neg(p \wedge \neg q \vee \neg r) \vee (\neg(q \wedge \neg p \vee r) \vee s)$$

Now, we apply De Morgan's laws to $\neg(p \wedge \neg q \vee \neg r)$ and $\neg(q \wedge \neg p \vee r)$ so that $\neg$ is only associated with atoms.

$$\varphi = \neg((p \wedge \neg q) \vee \neg r) \vee (\neg((q \wedge \neg p) \vee r) \vee s)$$

$$= (\neg(p \wedge \neg q) \wedge \neg\neg r) \vee ((\neg(q \wedge \neg p) \wedge \neg r) \vee s)$$

$$= ((\neg p \vee \neg\neg q) \wedge r) \vee (((\neg q \vee \neg\neg p) \wedge \neg r) \vee s)$$

$$= ((\neg p \vee q) \wedge r) \vee (((\neg q \vee p) \wedge \neg r) \vee s)$$

The last step is to distribute $\vee$ over $\wedge$ using the distributive law.

$$\varphi = ((\neg p \vee q) \wedge r) \vee (((\neg q \vee p) \vee s) \wedge (\neg r \vee s))$$

$$= ((\neg p \vee q) \wedge r) \vee ((\neg q \vee p \vee s) \wedge (\neg r \vee s))$$

$$= (((\neg p \vee q) \wedge r) \vee (\neg q \vee p \vee s)) \wedge (((\neg p \vee q) \wedge r) \vee (\neg r \vee s))$$

Let

$$\varphi_1 = ((\neg p \vee q) \wedge r) \vee (\neg q \vee p \vee s)$$

$$\varphi_2 = ((\neg p \vee q) \wedge r) \vee (\neg r \vee s)$$

Applying the distributive law to φ_1 and φ_2, we have

$$\varphi_1 = ((\neg p \vee q) \vee (\neg q \vee p \vee s)) \wedge (r \vee (\neg q \vee p \vee s))$$

$$= (\neg p \vee q \vee \neg q \vee p \vee s) \wedge (r \vee \neg q \vee p \vee s)$$

$$\varphi_2 = ((\neg p \vee q) \vee (\neg r \vee s)) \wedge (r \vee (\neg r \vee s))$$

$$= (\neg p \vee q \vee \neg r \vee s) \wedge (r \vee \neg r \vee s)$$

Thus, the formula in CNF is

$$\varphi = \varphi_1 \wedge \varphi_2$$

$$= (\neg p \vee q \vee \neg q \vee p \vee s) \wedge (r \vee \neg q \vee p \vee s) \wedge (\neg p \vee q \vee \neg r \vee s)$$

$$\wedge (r \vee \neg r \vee s)$$

φ in CNF has four clauses. There is a pair of q and $\neg q$ in the first clause, and a pair of r and $\neg r$ in the fourth clause, but there are no such negated pairs in the second and third clauses. Therefore, φ is not valid.

4.5.2 Horn Clauses

A formula is a *Horn formula* if it is in CNF and every conjunct contains at most one positive literal (not a negated atom). For example,

$$\varphi = (p \vee \neg q) \wedge (\neg p \vee \neg r \vee q) \wedge \neg s \wedge p$$

is a Horn formula. Each conjunct in a Horn formula is called a *Horn clause*. Because of the equivalence $\neg p \vee q \equiv p \rightarrow q$, we can convert each Horn clause to an implication. For example, for the Horn clauses in φ,

$$p \vee \neg q = q \rightarrow p$$

$$\neg p \vee \neg r \vee q = \neg(p \wedge r) \vee q = p \wedge r \rightarrow q$$

$$\neg s = \neg s \vee \bot = s \rightarrow \bot$$

$$p = \bot \vee p = \top \rightarrow p$$

Therefore, we can equivalently write φ as

$$\varphi = (q \rightarrow p) \wedge (p \wedge r \rightarrow q) \wedge (s \rightarrow \bot) \wedge (\top \rightarrow p)$$

Let

$$P ::= \bot \mid \top \mid p$$

where p represents an atom. Then, a Horn clause can also be defined as an implication whose assumption is a conjunction of atoms and whose conclusion is of type P. Examples of Horn formulas in which Horn clauses are of implication form are

$$(p \rightarrow q) \wedge (p \wedge q \rightarrow r)$$

$$(p \wedge q \wedge s \rightarrow r) \wedge (p \wedge q \rightarrow \bot) \wedge (\top \rightarrow p)$$

$$(r \rightarrow p) \wedge (\top \rightarrow s) \wedge (q \wedge s \wedge t \rightarrow p) \wedge (p \wedge q \wedge t \rightarrow \bot)$$

The following three formulas are not Horn formulas:

$$(p \rightarrow q) \wedge (\neg q \rightarrow r)$$

$$(p \wedge q \wedge s \rightarrow r \vee q) \wedge (p \wedge q \rightarrow \bot) \wedge (\top \rightarrow p)$$

$$(p \wedge q \vee s) \wedge (q \wedge s \wedge t \rightarrow p) \wedge (p \wedge q \wedge t \rightarrow \bot)$$

The first formula has a negative literal in the second clause; the second formula has a conclusion that is not an atom; and the first clause in the third formula is not an implication.

Given a Horn formula in which each Horn clause is in the implication form, we have an efficient algorithm to decide if the formula is satisfiable. The algorithm, Horn (φ), is described as follows:

```
Function Horn(φ):
/* precondition: φ is a Horn formula*/
/* postcondition: Horn(φ) decides satisfiability for φ */
Begin function
    mark all occurrences of p if ⊤ → p is a clause
    while there is a clause p₁ ∧ p₂ ∧ … pₖ → q such that all pᵢ
       are marked but q isn't do
          if q is ⊥ then return 'unsatisfiable'
          else mark q for all occurrences
    end while
    return 'satisfiable'
End function
```

The algorithm is explained as follows: as Horn formulas are a subclass of formulas in CNF, if a truth assignment satisfies a Horn formula, it must also satisfy each clause of the formula. As such, consider a Horn formula φ:

1. If $\top \to p$ is a clause in φ, then p must be true; otherwise, $\top \to p$ will evaluate to false and φ won't be satisfiable. Therefore, we mark p (as true).

2. By the same token, if $p_1 \wedge p_2 \wedge \ldots p_k \to q$ is a clause of φ and all p_i have already been marked, then q must also be marked, because it has to be true in order for the clause to be satisfiable.

3. If $p_1 \wedge p_2 \wedge \ldots p_k \to \perp$ is a clause of φ and all p_i have already been marked, then there is an inconsistency in the implication: the hypothesis is true while the conclusion is false. Therefore, the clause is false, and φ is not satisfiable.

4. Because a Horn clause of implication only evaluates to false in case 3 above, if such a case does not exist, then φ is satisfiable.

According to the algorithm, if either $\top$ or $\perp$ does not appear in φ, then φ is satisfiable. If there is $\top$ but no $\perp$, or neither $\top$ nor $\perp$, we can let all variables be true; If there is $\perp$ but no $\top$, we can let all variables be false.

It is worth pointing out that not all formulas can be converted equivalent Horn formulas.

Example 4.20: Satisfiability of Horn Formulas

Given the Horn formulas

a. $\varphi_1 = (p \wedge q \wedge s \rightarrow p) \wedge (q \wedge r \rightarrow p) \wedge (p \wedge s \rightarrow s)$
b. $\varphi_2 = (p \rightarrow s) \wedge (q \wedge r \rightarrow p) \wedge (r \wedge s \rightarrow \perp)$
c. $\varphi_3 = (\top \rightarrow q) \wedge (\top \rightarrow s) \wedge (q \wedge r \rightarrow p) \wedge (p \wedge q \wedge s \rightarrow \perp) \wedge (\top \rightarrow r)$
d. $\varphi_4 = (q \rightarrow p) \wedge (q \wedge r \wedge t \rightarrow \perp) \wedge (\top \rightarrow s) \wedge (s \rightarrow q)$

Decide their satisfiability.

1. Because there is no clause of form $\top \rightarrow p$, there is no initial marking of any atom. Therefore, the while-loop condition in the Horn algorithm is not true, and thus it is skipped. As a consequence, the algorithm returns "satisfiable." Actually, if we assign T's to all variables in the formula φ_1, φ_1 will be true, which means it is satisfiable.
2. Like in (a), there is no clause of form $\top \rightarrow p$ in φ_2 so φ_2 is satisfiable. Actually, if we assign F's to all variables in the formula φ_2, φ_2 is true.
3. The following Horn clauses

$$\top \rightarrow q$$

$$\top \rightarrow s$$

$$\top \rightarrow r$$

force q, s, and r to be true, respectively, and thus we mark q, s, and r in all clauses in the first round. After that, the clause

$$q \wedge r \rightarrow p$$

has q and r marked, but not p, and so we mark p in the second round. Now, we see the clause

$$p \wedge q \wedge s \rightarrow \perp$$

has all p, q and s marked, but the conclusion is a contradiction. According to the algorithm, φ_3 is not satisfiable.

4. For the formula φ_4, because of the clause $\top \rightarrow s$, we mark s in the first round. Then through $s \rightarrow q$ we mark q in the second round and through $q \rightarrow p$ we mark p in the third round. After that, there is no clause that satisfies the while-loop condition in the algorithm and thus φ_4 is satisfiable. One truth assignment satisfying φ_4 is all p, q, r, and s are true and t is false.

Exercises

1. For each of the following declarative sentences, identify atomic propositions, and write the sentence in propositional logic form using propositional logic operators.

   a. George is a college student and George plays football.

   b. If today is Monday, then tomorrow will be Tuesday.

   c. If today is Monday, then tomorrow will not be Friday.

   d. Tom is either from Florida or from Texas.

   e. If Mary ranks first in the class, she will receive the academic excellence award. She didn't receive the academic award, so she didn't rank first in the class.

   f. The professor is always late for class if there is a traffic jam. He is late, so there must be a traffic jam.

   g. If you study very hard, you will pass the class. If you do not study very hard, you may or may not pass the class.

   h. If you buy lottery tickets, you may win the lottery. If you do not buy lottery tickets, you will not win.

   i. If the traffic signal is neither green nor yellow, then it must be red.

   j. Every object is moving and meanwhile not moving.

   k. The light bulb can be at one of three states: on, off, and broken.

2. Construct truth tables for the following propositional logic formulas.

   a. $\neg(p \wedge \neg r)$

   b. $\neg p \rightarrow \neg q$

   c. $\neg p \wedge q \rightarrow r$

   d. $p \vee (\neg q \wedge r)$

   e. $\neg(p \wedge r) \vee q$

   f. $p \wedge (r \rightarrow q)$

   g. $p \wedge (r \rightarrow q) \rightarrow r$

   h. $p \rightarrow q \rightarrow r \rightarrow \neg p$

   i. $q \rightarrow p \vee r \rightarrow \neg q$

3. Show the following logic equivalencies using truth tables.

   a. $p \rightarrow q \equiv \neg q \rightarrow \neg p$

   b. $\neg(p \rightarrow q) \equiv p \wedge \neg q$

   c. $\neg(p \wedge q) \equiv \neg p \vee \neg q$

   d. $\neg(p \vee q) \equiv \neg p \wedge \neg q$

   e. $p \wedge (q \vee r) \equiv (p \wedge q) \vee (p \wedge r)$

   f. $p \vee (q \wedge r) \equiv (p \vee q) \wedge (p \vee r)$

4. Prove the validity of the following arguments with inference rules.

   a. $p \wedge q \vdash p \vee r$

   b. $p \wedge q, r \vdash q \wedge r$

   c. $p \wedge q, \neg r \vdash (\neg p \vee \neg r) \vee (p \vee \neg q)$

   d. $p \rightarrow q \vdash \neg q \rightarrow \neg r$

   e. $p, q \vdash (p \vee \neg r) \wedge (q \vee \neg r)$

   f. $(p \rightarrow q) \wedge (p \rightarrow r) \vdash p \rightarrow q \wedge r$

   g. $p \rightarrow q, q \rightarrow r, \neg r \vdash \neg p$

   h. $\neg q, \neg r, p \rightarrow r \vee q, \neg p \rightarrow t \vdash t$

   i. $p \rightarrow r \rightarrow q \vdash \neg q \rightarrow (p \wedge \neg r)$

   j. $\neg q \vdash (p \rightarrow q) \vee \neg p$

   k. $\vdash (p \rightarrow q) \rightarrow (p \rightarrow (q \vee r))$

   l. $p \rightarrow q, r \rightarrow \neg s \vdash p \wedge r \rightarrow q \wedge \neg s$

   m. $p \rightarrow q, r \rightarrow q, \neg q \vdash \neg(p \vee r)$

   n. $p \vee q, p \rightarrow (r \wedge \neg s), q \rightarrow s \vdash r \vee s$

   o. $p \vee (q \rightarrow p), q \vdash p$

5. Prove the following arguments using PBC.

   a. $\vdash p \vee \neg p$

   b. $\neg p \wedge \neg q \vdash \neg(p \vee q)$

   c. $p \vee q, \neg q \vdash p$

   d. $\neg \neg p \vdash p$

   e. $\neg(p \vee \neg q) \vdash q$

   f. $p \rightarrow q, \neg q \vdash \neg p$

   g. $p \rightarrow q \rightarrow r, q \wedge \neg r \vdash \neg p$

   h. $(p \vee q) \rightarrow (q \wedge r), \neg r \vdash \neg p$

6. Prove the following arguments using LEM.

   a. $p \rightarrow q, p \vee q \vdash p$

   b. $\vdash (p \rightarrow q) \vee \neg q$

   c. $p \rightarrow r, q \rightarrow p \vdash \neg q \vee r$

   d. $\vdash (p \rightarrow q) \vee (q \rightarrow r)$

   e. $p \vdash (p \wedge q) \vee (p \wedge \neg q)$

   f. $p \vee q, q \rightarrow r \vdash p \vee r$

7. Find out problems with the proofs of the following arguments.

a. $p \rightarrow q, q \vdash p \wedge q$

1.	$p \rightarrow q$	Premise
2.	q	premise
3.	p	MT, 1, 2
4.	$p \wedge q$	$\wedge i$, 2, 3

b. $r \rightarrow s \vdash r \rightarrow q \wedge s$

1.	$r \rightarrow s$	Premise
2.	r	Assumption
3.	s	MP, 1, 2
4.	$q \wedge s$	$\vee i$, 4
5.	$r \rightarrow q \wedge s$	$\rightarrow i$, 2–4

c. $p \rightarrow q, q \rightarrow r \vdash p \wedge r$

1.	$p \rightarrow q$	Premise
2.	$q \rightarrow r$	Premise
3.	p	Assumption
4.	q	MP, 1, 2
5.	r	MP, 2, 4
6.	$p \wedge r$	$\wedge i$, 3, 5

d. $p \rightarrow (q \vee r), q \rightarrow r \vdash p \rightarrow r$

1.	$p \rightarrow q \vee r$	Premise
2.	$q \rightarrow r$	Premise
3.	p	Assumption
4.	$q \vee r$	MP, 1, 4
5.	q	$\vee e$, 4
6.	r	MP, 2, 5
7.	$p \rightarrow r$	$\rightarrow i$, 3–5

e. $p \rightarrow q, q \rightarrow r, \neg r \vdash p \wedge r$

1.	$p \rightarrow q$	Premise
2.	$q \rightarrow r$	Premise
3.	$\neg r$	Premise
4.	p	Assumption
5.	q	MP, 1, 4
6.	r	MP, 2, 5
7.	$\perp$	Contradiction 3, 6
8.	p	PBC, 4–7
9.	$p \wedge r$	$\wedge i$, 8, 6

f. $p \to q, \neg r \vdash q \land \neg r$

   1. $p \to q$ Premise

   2. $\neg r$ Premise

   3. $p \lor \neg p$ LEM

   4. p Assumption

   5. q MP, 1, 4

   6. $q \land \neg r$ $\land i$, 2, 5

   7. $q \land \neg r$ $\lor e$, 3, 4–7

8. Convert the following formulas to CNF and check their validity.

   a. $p \lor q \land s \lor t$

   b. $p \to q \to r$

   c. $(p \land q) \to (p \land \neg q)$

   d. $p \to (q \lor \neg r) \land q \to r$

   e. $(p \lor q \to \neg r \land s) \lor (r \to q \land s)$

   f. $(p \to (q \land \neg r \land s)) \land (r \land q \to s)$

   g. $(p \to \neg q) \to (r \land q)$

   h. $(p \lor \neg r) \land q \to (r \land q \to s)$

9. Convert each of the following Horn formulas to an equivalent Horn formula in which each Horn clause is an implication.

   a. $p \land q \land r$

   b. $\neg p \land \neg q \land r$

   c. $p \land (\neg s \lor r) \land (\neg p \lor \neg r \lor q)$

   d. $(\neg p \lor \neg q) \land (\neg r \lor \neg s \lor p) \land q \land \neg s$

   e. $(\neg p \lor q) \land (\neg p \lor \neg r \lor \neg s \lor q) \land r \land s \land (\neg p \lor r) \land (\neg q \lor s)$

   f. $\neg p \land (\neg q \lor \neg r \lor p) \land r \land s \land (\neg s \lor q) \land (\neg p \lor \neg q)$

10. Apply the Horn(φ) algorithm to the following Horn formulas.

   a. $(p \land s \to r) \land (p \to \top) \land (p \to s)$

   b. $(p \to q) \land (p \land q \to r) \land (\top \to r)$

   c. $(\top \to q) \land (p \land q \to \bot) \land (\top \to p) \land (p \to q)$

   d. $(p \land q \to r) \land (q \land r \to \top) \land (q \to T) \land (q \to r) \land (q \to \bot)$

   e. $(p \to \top) \land (r \to \top) \land (q \land r \to s) \land (p \land r \to s)$

   f. $(p \land q \to s) \land (p \land r \to \top) \land (q \land r \to \bot)$

   g. $(p \to \top) \land (\top \to r) \land (q \land r \to \bot) \land (p \land r \to q)$

   h. $(p \land q \land s \land t \to r) \land (r \land s \to p) \land (r \to \top) \land (r \to s)$

   i. $(p \land q \to \top) \land (q \land r \land s \to p) \land (q \to \top) \land (p \land t \to r) \land (p \land q \to s) \land (p \land s \to \bot)$

5

Predicate Logic

Propositional logic is a declarative logic. It assumes the world contains *facts*. Propositional logic cannot identify "individuals," or specify relations between individuals. For example:

> Mike runs fast.
> Mike runs faster than John.

In these two statements Mike and John are individuals. We may also have statements like the following:

> Jack runs fast.
> Steve runs fast.
> Mary runs fast.
> ... ...

Structure-wise, they are similar to *Mike runs fast*. However, when we code them in propositional logic, we can only use p, q, r ... to represent them, which completely hides the similarity of these logic sentences.

Also consider a very common argument like the following:

> All dogs bark.
> Peter is a dog.
> Therefore, Peter barks.

This argument is valid, but the propositional logic system cannot thoroughly express the statements and prove the validity, because when you code the three atomic statements with letters like p, q, and r, there is no way to tell the relations among them. Therefore, propositional logic has very limited expressive power. We need a more powerful logic to deal with this limitation.

This chapter introduces *predicate logic*. Predicate logic, also called *first-order logic* and *first-order calculus*, assumes the world contains objects. Objects are represented with variables. The relations between objects are specified with predicates.

5.1 Predicates

A *predicate* is an expression involving one or more variables. It is a template that specifies a property of objects, or a relation among objects represented by variables. Substitution of a particular value for the variable(s) produces a proposition that is either true or false. For example, consider the mathematic assertion:

$$P(n): n \text{ is prime}$$

It is a predicate in which P is a *predicate symbol* and n is a variable. Substituting n with 2 we get

$$P(2): 2 \text{ is prime}$$

$P(2)$ is a proposition and is true. However, if we substitute n with 4 we get

$$P(4): 4 \text{ is prime}$$

The proposition $P(4)$ is obviously false.

A variable must be defined on a *domain*, which is also called the *universe of discourse*, or simply *universe*. The domain of a variable is the set of all values that may be substituted in place of the variable. The domain of n in the predicate $P(n)$, for example, is the natural numbers. In practice, domains of variables in predicates are often left implicit, but should be clear from the context.

Example 5.1: Predicates

Let x and y be integers and define

$$P(x, y): x + y = 10$$

Then $P(1, 1)$, $P(1, 2)$, $P(2, 1)$, and $P(2, 2)$ are all false, and $P(1, 9)$, $P(9, 1)$, $P(2, 8)$, and $P(8, 2)$ are all true.

Example 5.2: Use of Predicates

Define predicates

$S(x, y): x$ *is a student of* y *college*
$P(x): x$ *can program in Java*

Then the compound expression $S(x, y) \land P(x)$ states "x is a student of y college and x can program in Java," and the expression $S(x, y) \rightarrow P(x)$ states "if x is a student of y college, then x can program in Java."

Notice that the variables used in predicates are mere placeholders; their names are not important. However, we do need to use them consistently. For example, when we write $S(x, y) \rightarrow P(x)$, the x in both $S(x, y)$ and $P(x)$ is referring to the same object. If we write $S(x, y) \rightarrow P(z)$, then the x in $S(x, y)$ and z in $P(z)$ may point to different objects.

Suppose we know that Cindy and John are two students of Monmouth University and Cindy can program in Java but John can't. We use symbols c, j, and m to represent Cindy, John, and Monmouth University, respectively. Then the following propositional formulas are true:

$$S(c, m)$$

$$S(c, m) \wedge P(c)$$

$$S(c, m) \wedge S(j, m)$$

$$P(c) \wedge \neg P(c)$$

However, the following formulas are false:

$$S(j, m) \wedge P(j)$$

$$P(c) \rightarrow P(j)$$

5.2 Quantifiers

A quantifier is a word or a phrase that is used to modify a noun to indicate the amount or quantity. Examples are *one* apple, *five* pears, and *50* tomatoes. In logic, the two most common quantifiers mean "for all" and "there exists." The former one is called the *universal* quantifier, and the latter one is called the *existential* quantifier. The traditional symbol for the universal quantifier is ∀, a rotated letter A. The symbol for the existential quantifier is ∃, a rotated letter E.

5.2.1 Universal Quantifier

Universal quantifier is a quantifier that indicates the predicate within its scope is true for all values of any variable included in the quantifier. For example, using the predicate symbols S and P defined in Example 5.2, if we want to state "all students of y college can program in Java," we can use the universal quantify ∀ and specify it as follows:

$$\forall x \ (S(x, y) \rightarrow P(x))$$

A straightforward translation of the formula is "For any object x, if x is a student of y college, then x can program in Java." Here, the quantifier $\forall x$ applies to $S(x, y) \rightarrow P(x)$, and it is important to include the whole implication statement in the external parenthesis. It will be a mistake if we write $\forall x \, S(x, y) \rightarrow P(x)$ because here $\forall x$ only applies to $S(x, y)$.

Notice that the following statement

$$\forall x \, (S(x, y) \wedge P(x))$$

expresses that "all people are students at y college and they all can program in Java." It has a completely different meaning from $\forall x \, (S(x, y) \rightarrow P(x))$. Typically, $\rightarrow$ is the main connective with $\forall$.

If we want to express "every student in every college can program in Java," we can write

$$\forall x \forall y \, (S(x, y) \rightarrow P(x))$$

Example 5.3: Universal Quantifier

Define predicates

$F(x)$: *x plays football*
$G(x)$: *x plays guitar*

Translate the following statements to predicate logic:

1. Everyone who plays football also plays guitar.
2. All guitar players do not play football.
3. Everyone who plays football does not play guitar.
4. Mark plays football but he does not play guitar.

Solution

1. $\forall x \, (F(x) \rightarrow G(x))$
2. $\neg \forall x \, (G(x) \rightarrow F(x))$
3. $\forall x \, (F(x) \rightarrow \neg G(x))$
4. $F(\text{Mark}) \wedge \neg G(\text{Mark}))$

5.2.2 Existential Quantifier

Existential quantifier is a quantifier that indicates the predicate within its scope is true for some values of any variable included in the quantifier. For example, using the predicate symbols S and P defined in Example 5.2, the statement

$$\exists x \, (S(x, y) \wedge P(x))$$

says "some students at y college can program in Java." Typically, $\wedge$ is the main connective with $\exists$.

A common mistake is to use $\rightarrow$ as the main connective with $\exists$. Consider the following statement:

$$\exists x \, (S(x, y) \rightarrow P(x))$$

This statement is true if there is a student at y college and he can program in Java. However, the statement is also true if there is someone who is not a student of y college, because in that case $S(x, y)$ is false, and thus $S(x, y) \rightarrow P(x)$ is true. The is not what we intend to express.

Example 5.4: Existential Quantifier

Define predicates

$F(x)$: *x plays football*
$G(x)$: *x plays guitar*

Translate the following statements to predicate logic:

1. Some people play both football and guitar.
2. Some people play guitar but do not play football.
3. Nobody plays both football and guitar.

Solution

1. $\exists x \, (F(x) \wedge G(x))$
2. $\exists x \, (\neg F(x) \wedge G(x))$
3. $\neg \exists x \, (F(x) \wedge G(x))$

Example 5.5: Multiple Variables

Define predicates

$I(x)$: x is an integer
$N(x)$: x is a natural number
$G(x, y)$: x is greater than y

Translate the following statements to predicate logic:

1. Every natural number is an integer.
2. All integers are not natural numbers.
3. All natural numbers are greater than some integers.

Solution

1. $\forall x \, (N(x) \rightarrow I(x))$
2. $\neg \forall x \, (I(x) \rightarrow N(x))$
3. $\forall x \, \exists y \, (N(x) \rightarrow I(y) \wedge G(x, y))$

or

$$\forall x \, (N(x) \rightarrow \exists y \, (I(y) \wedge G(x, y)))$$

5.2.3 Properties of Quantifiers

Consider the predicate

$$\forall x\, P(x)\text{: } \textit{Everyone plays piano.}$$

How do we code the negation of the statement "not everyone plays piano?" A straightforward translation is

$$\neg \forall x\, P(x)\text{: } \textit{Not everyone plays piano.}$$

It reads "it is not true that everyone plays piano." This is equal to saying "there are some people who do not play piano," or "some people do not play piano," which can be coded as

$$\exists x\, \neg P(x)\text{: } \textit{Some people do not play piano.}$$

Therefore, we have an equivalency:

$$\neg\, \forall x\, P(x) \equiv \exists x\, \neg P(x)$$

With the same analysis, we can also find

$$\neg \exists x\, P(x) \equiv \forall x\, \neg P(x)$$

For example, the following two statements are equal:

There is not a man who can live forever.
Everyone cannot live forever.

Example 5.6: Negation of Quantifiers

Define

 $P(x)$: *x is a computer program*
 $V(x)$: *x is a computer virus*
 $K(x, y)$: *x kills y*

Then the formula

$$\exists x \forall y\, (P(x) \wedge V(y) \rightarrow K(x, y))$$

says

 There is a computer program that kills all viruses.

Find the negation

$$\neg(\exists x \forall y\, (P(x) \wedge V(y) \rightarrow K(x, y)))$$

Solution

$$\neg(\exists x \forall y \, (P(x) \wedge V(y) \rightarrow K(x, y)))$$

$$= \forall x(\neg \forall y \, (P(x) \wedge V(y) \rightarrow K(x, y)))$$

$$= \forall x \exists y(\neg(P(x) \wedge V(y) \rightarrow K(x, y)))$$

$$= \forall x \exists y(\neg(\neg(P(x) \wedge V(y)) \vee K(x, y)))$$

$$= \forall x \exists y(\neg\neg(P(x) \wedge V(y)) \wedge \neg K(x, y)))$$

$$= \forall x \exists y(P(x) \wedge V(y) \wedge \neg K(x, y))$$

It says

For every computer program there is a virus that it cannot kill.

The positions of the same type of quantifiers can be switched without affecting the truth value as long as there is no other type of quantifier in between the ones to be switched. For example, define

$Q(x, y)$: Rabbit x loves vegetable y.

Then,

$\forall x \forall y \, Q(x, y)$: All rabbits love all vegetables.
$\forall y \forall x \, Q(x, y)$: All vegetables are loved by all rabbits.

They are the same. Similarly,

$\exists x \exists y \, Q(x, y)$: Some rabbits love some vegetables.
$\exists y \exists x \, Q(x, y)$: Some vegetables are loved by some rabbits.

They are also the same.

However, we cannot switch the positions of two quantifiers of different types. Consider

$R(x, y)$: Student x respects professor y.
$\exists x \forall y \, R(x, y)$: There is a student who respects every professor.
$\forall y \exists x \, R(x, y)$: Every professor is respected by some student.

Obviously, the two formulas are different. To further illustrate the difference, let x, y be variables in the domain of natural numbers, and $G(x, y)$ mean x is greater than y. Then, $\forall y \exists x \, G(x, y)$ expresses

For any natural number, there exists a number that is greater than it.

We know mathematically this statement is correct. However, $\exists x \forall y\ G(x, y)$ expresses

There exists a natural number that is greater than any other numbers.

This statement is false.

5.3 Syntax of Predicate Logic

There are two key types of legal expressions in predicate logic: *terms* and *formulas*. Terms are objects. They are the individuals we refer to in the predicate, such as students, professors, rabbits, and vegetables. Formulas express predicates that can be evaluated to true or false.

5.3.1 Terms

Terms are formally defined as follows:

1. Any variable is a term.
2. Any constant (i.e., nullary function) is a term.
3. If $t_1, t_2, \ldots, t_n$ are terms and f is a n-ary function, then $f(t_1, t_2, \ldots, t_n)$ is a term.
4. Nothing else is a term.

Using BNF notation, the definition can be written as

$$t ::= x \,|\, c \,|\, f(t, t, \ldots, t)$$

where x is member of a set of variables, c ranges over a set of nullary function symbols, and f over a set of function symbols with at least one argument.

Function symbols are often denoted by lowercase letters f, g, h, etc. For example, we can use $f(x)$ to represent the father of x, and $f(f(x))$ to represent the grandpa of x. A function with 0 or more arguments (terms) evaluates to another term. A function with 0 terms is a constant. This is different from a predicate. A predicate can also have one or more variables, but it always evaluates to true or false. A predicate with 0 arguments is a proposition, such as p and q in the propositional logic.

Example 5.7: Functions as Terms

Consider the statement

Every child is younger than its mother.

Specify it with predicate logic expression.

Solution

Define predicates:

> $C(x)$: x is a child
> $M(x, y)$: x is the mother of y
> $Y(x, y)$: x is younger than y

We can translate the statement to

$$\forall x \forall y \, (C(x) \wedge M(y, x) \to Y(x, y))$$

This seems okay, but is really awkward because the component $\forall x \forall y M(y, x)$ reads "for any child and for any mother," but there is only one mother for each child. If we define a function

$$m(x): \text{the mother of } x,$$

then we can rewrite the logic expression as

$$\forall x \, (C(x) \to Y(x, m(x)))$$

The new expression reads much more nature. In $Y(x, m(x))$, both x and $m(x)$ are terms.

5.3.2 Formulas

A well-formed formula in predicate logic is inductively defined as

1. If P is an n-ary predicate symbol and $t_1, \ldots, t_n$ are terms, then $P(t_1, \ldots, t_n)$ is a formula.
2. If the equality symbol ($=$) is considered part of logic, and t_1 and t_2 are terms, then $t_1 = t_2$ is a formula.
3. If φ is a formula, then so is $(\neg\varphi)$.
4. If φ and ψ are formulas, then so are $(\varphi \wedge \psi)$, $(\varphi \vee \psi)$ and $(\varphi \to \psi)$.
5. If φ is a formula and x is a variable, then $(\forall x \, \varphi)$ and $(\exists x \, \varphi)$ are formulas.

In BNF notation, the definition can be written as

$$\Phi ::= P(t, t, \ldots, t) \mid (t = t) \mid \neg\varphi \mid (\varphi \wedge \varphi) \mid (\varphi \vee \varphi) \mid (\varphi \to \varphi) \mid (\forall x \, \varphi) \mid (\exists x \, \varphi)$$

where P is a predicate symbol and t is a term.

The equality symbol is a logic symbol that is only used between two terms. The equality $t_1 = t_2$ is true under some interpretation if and only if t_1 and t_2 refer to the same object. The symbol cannot be used between two formulas.

The formulas obtained from the first two rules are called *atomic formulas*. Let P, Q, R, S, and T be predicate symbols; x, y, and z be variables; f and g be function symbols; and a, b, c, and d be constants. The formulas that follow are all well-formed atomic formulas:

1. P
2. $P(x, y, z)$
3. $f(a, b) = g(c, d)$

The following are examples of well-formed predicate logic formulas:

1. $P(x, y, z) \wedge Q(x, y, z)$
2. $\forall x \forall y \exists z \, (P(x, y, z) \wedge Q(x, y, z))$
3. $\forall x \exists y \, (P(x, y) \wedge Q(x, y) \rightarrow S(x) \vee T(y)) \rightarrow \neg \exists z \, R(z)$

The following are examples of invalid predicate logic formulas:

1. x
2. $f(a, b)$
3. $P(x, y, z) \wedge Q(x, y, z) \rightarrow g(c, d)$
4. $\forall x \forall y \exists z \, (P(x, y, z) \rightarrow Q(x, y, R(z)))$
5. $\neg \exists z \, R(z) \, \forall x \exists y \, (P(x, y) \wedge Q(x, y) \rightarrow S(x) \vee T(y))$

Here formulas 1 and 2 are invalid because a term or function is not a formula. Formula 3 is invalid because $g(c, d)$ is a term and it does not have a logic value. Formula 4 is invalid because a predicate cannot be an argument of another predicate or function. Formula 5 is invalid because there must be a logic operator between any two (quantified) predicates.

Operators, connectives, and quantifiers in predicate logic are evaluated according to the following conventions: $\neg$ is evaluated first, $\forall x$ and $\exists y$ are evaluated next, $\wedge$ and $\vee$ are evaluated after, and $\rightarrow$ is evaluated last. Following these binding rules, some parentheses can be omitted in formulas. For example, the formula

$$\forall x \exists y \, ((P(x) \wedge Q(x, y)) \rightarrow (S(x) \vee T(y))) \rightarrow \neg(\exists z \, R(z))$$

can be simplified as

$$\forall x \exists y \, (P(x) \wedge Q(x, y) \rightarrow S(x) \vee T(y)) \rightarrow \neg \exists z \, R(z)$$

But oftentimes, we add extra grouping symbols to formulas to improve readability.

Example 5.8: Predicate Logic Formula

Translate the following sentence into predicate logic formula:

Bob and Paul have the same maternal grandmother.

Solution

We can solve the problem in several different ways. We can define a predicate

$$M(x, y): x \text{ is } y\text{'s maternal mother.}$$

Then Bob's grandmother mother, x, satisfies

$$\forall x \forall y \, (M(x, y) \wedge M(y, b))$$

where b is a constant for Bob. Similarly, Paul's grandmother, u, satisfies

$$\forall u \forall v \, (M(u, v) \wedge M(v, p))$$

where p is a constant for Paul. Because Bob and Paul have the same maternal grandmother, so x and u must be the same object. Therefore, the sentence is translated as

$$\forall x \forall y \forall u \forall v \, (M(x, y) \wedge M(y, b) \wedge M(u, v) \wedge M(v, p) \rightarrow x = u)$$

We can simplify the translation significantly if we introduce a function

$$m(x): \text{maternal mother of } x$$

With $m(x)$ being x's mother, x's grandmother can be expressed as $m(m(x))$, and thus the sentence can be coded as

$$m(m(b)) = m(m(p))$$

This translation is not only simplified, but also reads much more elegantly than the previous one.

5.3.3 Parse Trees

We can draw a parse tree for each well-formed predicate logic formula, just like for each propositional logic formula. There are three new types of nodes in predicate logic parse trees, though. They are

- The quantifiers $\forall x$ and $\exists y$ form nodes that have one subtree.
- Any predicate of n-ary $P(t_1, t_2, \ldots, t_n)$ has the symbol P as a node and the node has a subtree for each of the terms $t_1, t_2, \ldots, t_n$.
- Any function of n-ary $f(t_1, t_2, \ldots, t_n)$ has the symbol f as a node, and the node has a subtree for each of the terms $t_1, t_2, \ldots, t_n$.

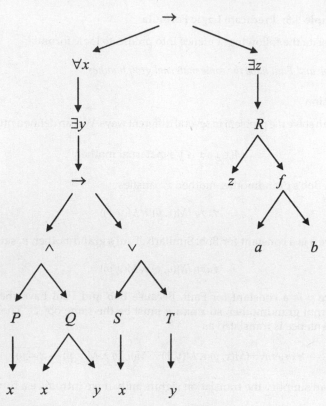

FIGURE 5.1
Parse tree of $\forall x \exists y \, (P(x) \wedge Q(x, y) \rightarrow S(x) \vee T(y)) \rightarrow \exists z \, R(z, f(a, b))$.

Figure 5.1 shows the parse tree of the formula $\forall x \exists y \, (P(x) \wedge Q(x, y) \rightarrow S(x) \vee T(y)) \rightarrow \exists z \, R(z, f(a, b))$. Each subtree represents a subformula. For example, the left subtree of the root $\rightarrow$ represents the subformula

$$\forall x \exists y \, (P(x) \wedge Q(x, y) \rightarrow S(x) \vee T(y))$$

and the right subtree represents

$$\exists z \, R(z, f(a, b))$$

5.3.4 Free and Bound Variables

All variables that are not quantified in a formula are *free* variables, and those quantified are *bound* variables. Consider the formula

$$\forall x (P(x) \wedge Q(x, y))$$

The scope of $\forall x$ is $(P(x) \wedge Q(x, y))$ and thus both of the two occurrences of x are bound. y is free though. The free and bound variables are defined inductively as follows:

1. If φ is an atomic formula, then x is free in φ if and only if x occurs in φ. Moreover, there are no bound variables in any atomic formula.
2. x is free in $\neg\varphi$ if and only if x is free in φ. x is bound in $\neg\varphi$ if and only if x is bound in φ.
3. x occurs free in $(\varphi \wedge \psi)$ if and only if x occurs free in either φ or ψ. x occurs bound in $(\varphi \wedge \psi)$ if and only if x occurs bound in either φ or ψ. The same rule applies to other two binary connectives.
4. x is free in $\forall y\, \varphi$ if and only if x is free in φ and x is a different symbol free from y. Also, x is bound in $\forall y\, \varphi$ if and only if x is y or x is bound in φ. The same rule applies to the case with $\exists$ in place of $\forall$.

For example, consider this formula:

$$P(x) \wedge \forall x Q(x, y) \rightarrow \neg\exists z R(z)$$

According to 1, x in $P(x)$ is free. According to 4, y in $\forall x\, Q(x, y)$ is free and x in $\forall x\, Q(x, y)$ is bound. Also according to 4, z in $\exists z R(z)$ is bound. Applying the rules in 2, z in $\neg\exists z R(z)$ is bound. According to 3, y in $P(x) \wedge \forall x Q(x, y)$ is free. Moreover, the first occurrence of x in $P(x) \wedge \forall x Q(x, y)$ is free but the second occurrence is bound.

Notice that this example also shows that in a formula free variables and bound variables do not have to be disjoint. In fact, the x in $P(x)$ and x in $\forall x Q(x, y)$ are just placeholders. They may have nothing to do with each other at all.

Free and bound variables can be easily determined through parse trees. Choose a leaf node for an occurrence of a variable and walk up the tree. If a quantifier for the variable is encountered, then this occurrence of the variable is bound. If no quantifiers for the variable are hit all the way until the root is reached, then the occurrence of the variable is free. Consider

$$\forall x\, (P(x) \wedge Q(x, y) \rightarrow S(x) \vee T(y)) \rightarrow R(z, f(a, b))$$

Its parse tree is shown in Figure 5.2. From left to right, the first three occurrences of x are all underneath $\forall x$, and thus they are bound. There are no x related quantifiers from the root to the fourth occurrence of x. Therefore, the fourth occurrence of x is free. Since there are no quantifiers on y and z on the tree, they are all free.

A well-formed predicate logic formula that contains no free variables is a *sentence* because it can always be evaluated to a fixed truth value with a given model. On the other hand, the truth value of a formula with free variables may vary with the value assignment of its free variables. This will be discussed in Section 5.5.

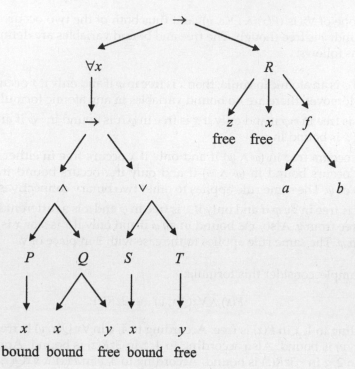

FIGURE 5.2
Free variables and bound variables.

5.3.5 Substitution

Let φ be a formula, x a variable, and t a term.

Replace each free occurrence of x in φ with t and denote the new formula by $\varphi[t/x]$.

For example, consider

$$\varphi: P(x) \wedge \forall x Q(x, y) \rightarrow \neg \exists z R(z)$$

Substituting free occurrences of x with t will give

$$\varphi[t/x]: P(t) \wedge \forall x Q(x, y) \rightarrow \neg \exists z R(z)$$

Notice that only the first occurrence of x in φ is free. Therefore, it is changed to t. The second occurrence of x is bound, and thus there is no change to it.

If we substitute y in φ with $f(a, b)$, we will get

$$\varphi[f(a, b)/y]: P(x) \wedge \forall x Q(x, f(a, b)) \rightarrow \neg \exists z R(z)$$

What if we substitute z in φ with $f(a, b)$? Well, there will be no change because there is only one occurrence of z in φ and it is bound.

Can we substitute y in φ with x? If we do that, we will get

$$\varphi[x/y]: P(x) \wedge \forall x Q(x, x) \rightarrow \neg \exists z R(z)$$

The free variable y now becomes bound. Let us look at another example. Consider

$$\varphi : \forall x \, (x = y)$$

If we substitute y with $x + 1$. We get

$$\varphi[x + 1/y]: \forall x \, (x = x + 1)$$

Clearly, we've run into a problem. A more general issue is, in performing a substitution $\varphi[t/x]$, the term t may contain a variable y, which is quantified by $\exists y$ or $\forall y$. A free occurrence of x in φ is under the scope of $\exists y$ or $\forall y$. The variable y in the term t may have a specific value before the substitution. After the substitution, however, it stands for "some unspecified" or "all." This is called *variable capture*.

We should avoid variable capture. Given a term t, a variable x, and a formula φ, we say that t *is free for x in φ* if no free x in φ occurs in the scope of $\forall y$ or $\exists y$ for any variable y occurring in t. It is safe to substitute t for x in φ if t is free for x in φ. Otherwise, variable capture arises.

Variable capture can be avoided by a different choice of variable. For example, for $\varphi = \forall x \, (x = y)$, before we perform $\varphi[x/y]$, we change x to z and get $\forall z \, (z = y)$, which is equal to φ. After that, we do $\varphi[x/y]$ and get $\varphi: \forall z \, (z = x)$.

Now let us give a formal definition of substitution. Let x be a variable and t a term. For a term u, the term $[t/x]$ is u, with each occurrence of the variable x replaced by the term t. For a formula φ:

1. If φ is $P(t_1, t_2, \ldots, t_k)$, then $\varphi[t/x]$ is $P(t_1[t/x], t_2[t/x], \ldots, t_k[t/x])$.
2. If φ is $(\neg \varphi_1)$, then $\varphi[t/x]$ is $(\neg \varphi_1[t/x])$.
3. If φ is $(\varphi_1 \wedge \varphi_2)$, then $\varphi[t/x]$ is $(\varphi_1[t/x] \wedge \varphi_2[t/x])$. This also applies to $(\varphi_1 \vee \varphi_2)$ and $(\varphi_1 \rightarrow \varphi_2)$
4. If φ is $(\forall x \, \varphi_1)$ or $(\exists x \, \varphi_1)$, then $\varphi[t/x]$ is φ.
5. If φ is $(\forall y \, \varphi_1)$ or $(\exists y \, \varphi_1)$, then
   a. If y does not occur in t, then $\varphi[t/x]$ is $(\forall y \, \varphi_1[t/x])$ or $(\exists y \, \varphi_1[t/x])$, respectively.
   b. Otherwise, we select a variable z that occurs in neither φ nor t to replace *all* y first, and then we substitute x with t.

The last case prevents variable capture by renaming the quantified variable to something harmless.

Example 5.9: Substitution

Given

1. φ_1: $\forall x\ (P(x) \rightarrow Q(y))$
2. φ_2: $\forall x\ (P(x, y)) \rightarrow \exists y\ (Q(y))$
3. φ_3: $P(x) \wedge \exists x \forall y\ (Q(x, y)) \rightarrow \forall y\ (R(y))$
4. φ_4: $\forall x \exists y\ (P(x, y) \wedge Q(x)) \rightarrow \exists x\ (R(x, y))$

Find $\varphi_1[c/y]$, $\varphi_2[c/y]$, $\varphi_3[x/y]$, and $\varphi_4[(x + c)/y]$, where c is a constant.

Solution

1. The only occurrence of y is free, and thus it is replaced with c:

$$\varphi_1[c/y]:\ \forall x\ (P(x) \rightarrow Q(c))$$

2. There are two occurrences of y in the formula. The first one is free and substituted. The second occurrence is bound by $\exists y$ and is not substituted. Therefore,

$$\varphi_2[c/y]:\ \forall x\ (P(x, c)) \rightarrow \exists y\ (Q(y))$$

3. In this formula, there is no free y that occurs in the scope of $\exists x$ or $\forall x$. Thus x is free for y. Therefore,

$$\varphi_3[x/y]:\ P(y) \wedge \exists x \forall y\ (Q(x, y)) \rightarrow \forall y\ (R(y))$$

4. The term $t\ (= x + c)$ contains a variable x in this problem. We need to know whether x is free for y. We see the free y in $R(y)$ is in the scope of $\exists x$, and thus x is not free for y. To resolve the issue, we replace all x in $\exists x\ (R(x, y))$ with a different variable z first, which gives an equivalent φ_4 as

$$\varphi_4:\ \exists x\ (R(x, y)):\ \forall x \exists y\ (P(x, y) \wedge Q(x)) \rightarrow \exists z\ (R(z, y))$$

Then, we perform $\varphi_4[(x + c)/y]$:

$$\varphi_4[(x + c)/y]:\ \forall x \exists y\ (P(x, y) \wedge Q(x)) \rightarrow \exists z\ (R(z, x + c))$$

5.4 Natural Deduction Rules

The natural deduction rules that we introduced in Chapter 4 for propositional arguments are still valid in predicate logic systems. Because of the introduction of the two quantifiers and equality operator, additional deduction rules are added in predicate logic. This section describes them.

5.4.1 Rules for Equality

There are two proof rules for equality. The first rule is *equality introduction*. This rule simply states that each term is same as itself. The rule is written

$$\frac{}{t = t} = i$$

where t is a term. It does not have any precondition and is an axiom.

The equality introduction rule is seldomly used alone. Instead, it is often used together with the second equality rule, which is *equality elimination*:

$$\frac{t_1 = t_2, \varphi[t_1/x]}{\varphi[t_2 / x]} = e$$

It says given that the two terms t_1 and t_2 are the same, we may substitute the term t_2 for t_1 in φ.

Note that we do substitution $[t/x]$ in φ only if t is free x in φ. Without further statement, we always assume this is true in this section.

Example 5.10: The Sequent Expressing Symmetry

Show

$$t_1 = t_2 \vdash t_2 = t_1$$

Proof:

Define

$$\varphi[x]: x = t_1$$

Then

$$\varphi[t_1/x]: t_1 = t_1$$

From the equality introduction rule, we know this is always true. From the premise, we have

$$t_1 = t_2$$

From $t_1 = t_2$ and $\varphi[t_1/x]$, we know $\varphi[t_2/x]$ is also true, which is

$$\varphi[t_2/x]: t_2 = t_1$$

Therefore,

$$t_1 = t_2 \vdash t_2 = t_1$$

This proof process used both rules for equality. The definition of $\varphi[x]$ is to utilize the equality introduction rule. We follow the convention in Chapter 4 to rewrite the proof process as follows:

1. $t_1 = t_2$ Premise
2. $t_1 = t_1$ $= i$
3. $t_2 = t_1$ $= e, 1, 2$

Example 5.11: The Sequent Expressing Transitivity

Show

$$t_1 = t_2, t_2 = t_3 \vdash t_1 = t_3$$

Proof:

There are two premises in this sequent. If we use $t_2 = t_3$ as $\varphi[t_2/x]$, then $\varphi[t_1/x]$ will be the conclusion that we want to prove. Because $t_1 = t_2$, then according to the equality elimination rule, $\varphi[t_1/x]$ is true. The proof process is formally written as

1. $t_1 = t_2$ Premise
2. $t_2 = t_3$ Premise
3. $t_1 = t_3$ $= e, 1, 2$

5.4.2 Rules for Universal Quantifier

There are two proof rules for the universal quantifier. One is $\forall x$ introduction, and the other is $\forall x$ elimination.

The $\forall x$ elimination rule states that if $\forall x\ \varphi$ is true, then you can replace x in φ with any specific term t and conclude the result $\varphi[t/x]$ is true as well. This is similar to the conjunction elimination rule in propositional logic, in which we eliminate the conjunction by keeping only one conjunct and the conjunct is true. Notation-wise, $\forall$ elimination rule is written as

$$\frac{\forall x\ \varphi}{\varphi[t / x]}\ \forall x\ e$$

Example 5.12: $\forall x$ Elimination

Prove the argument

All dogs bark.
Peter is a dog.
Therefore, Peter barks.

Proof:

This is the argument we brought up at the beginning of this chapter. We pointed out that an argument like this cannot be proved with propositional logic because it cannot express individual objects and quantities. The proof, however, is fairly easy in predicate logic. Let us define the following predicates:

$D(x)$: x is a dog
$B(x)$: x barks

and also use p for Peter. Then the problem becomes to prove the following sequent:

$$\forall x \, (D(x) \rightarrow B(x)), D(p) \vdash B(p)$$

In this sequent, the conclusion is $B(p)$, a proposition. The predicate $B(x)$ appears in the first premise in the form $\forall x \, \varphi$. To get $B(p)$, we should apply the $\forall x \, e$ rule. The formal proof is as follows:

1. $\forall x \, (D(x) \rightarrow B(x))$ premise
2. $D(p)$ premise
3. $D(p) \rightarrow B(p)$ $\forall x \, e$, 1
4. $B(p)$ MP, 3, 2

To enclose φ in the scope of $\forall x$, we must prove φ being quantified for all possible values of x. Obviously, it is hard to prove it directly when the universe is large. We can, however, think in a different way: if we can prove φ is true for an arbitrarily selected fresh x in the universe, then we can claim φ is true for all x. In other words, we have $\forall x \, \varphi$. This is the $\forall x$ introduction rule, and it is written as

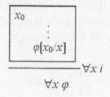

Notice that the scope of x_0 is limited to the proof box. When we say it is *fresh*, we mean that x_0 does not appear before we make the proof and will not be used after the proof. This is to ensure x_0 is a randomly selected value of x in its universe.

Example 5.13: $\forall x$ Introduction

Prove sequent

$$\forall x \, (P(x) \rightarrow Q(x)), \forall x \, P(x) \vdash \forall x \, Q(x)$$

Proof:

We want to prove $\forall x \, Q(x)$, which is in the form $\forall x \, \varphi$, and thus we can try to use the $\forall x \, i$ rule. We start the proof with the two premises. The proof structure should be like

$$\forall x \, (P(x) \to Q(x)) \qquad \text{Premise}$$
$$\forall x \, P(x) \qquad\qquad \text{Premise}$$

$$\begin{array}{|l|} \hline x_0 \\ \\ \vdots \\ \\ Q(x_0) \\ \hline \end{array}$$

$$\forall x \, Q(x)$$

We need to fill up the two potential gaps. Because both premises are in the form $\forall x \, \varphi$, we can apply the $\forall x \, e$ rule to get $P(x_0) \to Q(x_0)$ and $P(x_0)$ for any x_0. With these two formulas, we can apply the MP rule to get $Q(x_0)$. Because x_0 is an arbitrarily value of x, we can conclude $\forall x \, Q(x)$ according to the $\forall x \, i$ rule. The formal proof process is as follows:

1. $\forall x \, (P(x) \to Q(x))$		Premise
2. $\forall x \, P(x)$		Premise
3.	$x_0 \quad P(x_0) \to Q(x_0)$	$\forall x \, e \; 1$
4.	$P(x_0)$	$\forall x \, e \; 2$
5.	$Q(x_0)$	MP 3, 4
6. $\forall x \, Q(x)$		$\forall x \, i \; 3\text{-}5$

5.4.3 Rules for the Existential Quantifier

There are two proof rules for the existential quantifier: $\exists x$ introduction and $\exists x$ elimination. The $\exists x$ introduction rule states that if $\varphi[t/x]$ is true for a term t, then $\exists x \, \varphi(x)$ is true. This rule is fairly straightforward, and is written as

$$\frac{\varphi[t \, / \, x]}{\exists x \, \varphi(x)} \; \exists x \, i$$

We can compare this rule with $\vee \, i$: given φ and ψ, as long as one of them is true, we conclude $\varphi \vee \psi$ is true. In $\exists x \, i$, we are talking about a set of objects instead of two formulas. As long as one object from the set satisfies the formula, we can claim $\exists x \, \varphi(x)$. For example, the following sequent is valid:

For each student, if his semester GPA is 3.5 or better, he will be in the dean's list. ($\varphi[x]$)

Some students earned at least 3.5 semester GPA. ($\varphi[t/x]$)

Therefore, some students will be in the dean's list. ($\exists x \, \varphi(x)$)

The $\exists x$ elimination rule states that given a formula $\exists x \, \varphi(x)$ and an x_0, if by assuming $\varphi(x_0)$ is true, we can prove a formula χ, then we claim χ is true. The rule is written as

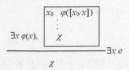

$$\exists x\, \varphi(x), \qquad \boxed{\begin{array}{l} x_0 \quad \varphi([x_0/x]) \\ \;\;\vdots \\ \quad \chi \end{array}}$$
$$\rule{5cm}{0.4pt}\; \exists x\, e$$
$$\chi$$

Recall the disjunction elimination rule. Given $\varphi \vee \psi$, in order to prove a third formula χ, we open an assumption box for φ and prove χ, and then open an assumption box for ψ and prove χ. If χ is proved for both cases, then we conclude χ is true. The $\exists x\, e$ rule deals with a set of objects. That $\exists x\, \varphi(x)$ is true means there is at least one object, say x_0, such that $\varphi(x_0)$ is true. If from $\varphi(x_0)$ being true we can prove that a formula, say χ, which has nothing to do with x_0, is true, then regardless of which individual x in the set satisfies $\varphi(x)$, we have χ.

Example 5.14: $\exists x$ Introduction

Show

$$\forall x\, (P(x) \rightarrow Q(x)),\ \exists x\, (P(x) \wedge R(x))\ \vdash\ \exists x\, (Q(x) \wedge R(x))$$

Proof:

The conclusion is in the form $\exists x\, \varphi$. There is also a premise in this form. So, we can try to use the $\exists x\, e$ rule to prove the conclusion. The structure of the proof should look like this:

$$\forall x\, (P(x) \rightarrow Q(x)) \qquad \text{Premise}$$
$$\exists x\, (P(x) \wedge R(x)) \qquad \text{Premise}$$

$$\boxed{\begin{array}{ll} x_0 \quad P(x_0) \wedge R(x_0) & \text{Assumption} \\ \quad\ \vdots \\ \quad \exists x\, (Q(x) \wedge R(x)) \end{array}}$$

$$\exists x\, (Q(x) \wedge R(x)) \qquad \exists x\, e$$

To fill the proof gap inside the assumption box, we should utilize the first premise. Because the first premise is in the form $\forall x\, \varphi$, we can apply the $\forall x\, e$ rule to it, and then work toward $\exists x\, (Q(x) \wedge R(x))$, the last formula inside the assumption box. The final proof process is listed as follows:

1.		$\forall x\, (P(x) \rightarrow Q(x))$	Premise
2.		$\exists x\, (P(x) \wedge R(x))$	Premise
3.	x_0	$P(x_0) \wedge R(x_0)$	Assumption
4.		$P(x_0) \rightarrow Q(x_0)$	$\forall x\, e,\ 1$
5.		$P(x_0)$	$\wedge\, e,\ 1, 3$
6.		$R(x_0)$	$\wedge\, e,\ 2, 3$
7.		$Q(x_0)$	MP, 4, 5
8.		$Q(x_0) \wedge R(x_0)$	$\wedge\, i,\ 7, 6$
9.		$\exists x\, (Q(x) \wedge R(x))$	$\exists x\, i,\ 8$
10		$\exists x\, (Q(x) \wedge R(x))$	$\exists x\, e,\ 2,\ 3\text{-}9$

Example 5.15: $\exists x$ Introduction and PBC

Show

$$\forall x\,(P(x) \vee Q(x)) \vdash \forall x\,P(x) \vee \exists x\,Q(x)$$

Proof:

The conclusion is a disjunction and thus as the first thought, we might want to prove either $\forall x\,P(x)$ is true or $\exists x\,Q(x)$ is true. To prove $\forall x\,P(x)$ is true, we can try to use the $\forall x\,i$ rule; to prove $\exists x\,Q(x)$ is true, we can try the $\exists x\,i$ rule. However, after a few tries you will find from the premise you can prove neither $\forall x\,P(x)$ nor $\exists x\,Q(x)$. Actually, you can find counter-examples that show neither of them is always true. On the other hand, there is no direct rule that allows us to prove $\forall x\,P(x) \vee \exists x\,Q(x)$ as a whole.

Recall we mentioned in Chapter 4 that when it is difficult to prove a conclusion directly from premises, we can try the rule of PBC, proof by contradiction. For this problem, the conclusion is $\forall x\,P(x) \vee \exists x\,Q(x)$, and its negation is

$$\neg(\forall x\,P(x) \vee \exists x\,Q(x))$$

The rough proof structure with PBC is as follows:

$\forall x\,(P(x) \vee Q(x))$	Premise
$\neg(\forall x\,P(x) \vee \exists x\,Q(x))$	Assumption
$\vdots$	
$\bot$	
$\forall x\,P(x) \vee \exists x\,Q(x)$	PBC

To fill the proof gap, we need to utilize both of the premise and the assumption. We cannot use the assumption directly, and thus we convert it to an equivalent:

$$\neg\forall x\,P(x) \wedge \neg\exists x\,Q(x)$$

Because we have no proof rules to deal directly with $\neg\forall x\,P(x)$ or $\neg\exists x\,Q(x)$, it is necessary to further convert them to $\exists x\,\neg P(x)$ and $\forall x\,\neg Q(x)$, respectively. With $\exists x\,\neg P(x)$, we can attempt to utilize the $\exists x\,e$ rule to prove the contradiction:

$\forall x\,(P(x) \vee Q(x))$	
$\neg(\forall x\,P(x) \vee \exists x\,Q(x))$	Assumption
$\neg\forall x\,P(x) \wedge \neg\exists x\,Q(x)$	Equivalent
$\neg\forall x\,P(x)$	$\wedge e$
$\neg\exists x\,Q(x)$	$\wedge e$
$\exists x\,\neg P(x)$	Equivalent
$\forall x\,\neg Q(x)$	Equivalent
$x_0 \quad \neg P(x_0)$	
$\vdots$	
$\bot$	
$\bot$	$\exists x\,e$
$\forall x\,P(x) \vee \exists x\,Q(x)$	PBC

Next, we should utilize the premise $\forall x\ (P(x) \vee Q(x))$ and formula $\forall x$ $\neg Q(x)$. They are all in the form $\forall x\ \varphi$, and thus we should apply the $\forall x\ e$ rule to get $P(x_0) \vee Q(x_0)$ and $\neg Q(x_0)$. Because $P(x_0) \vee Q(x_0)$ is a disjunction, we should follow the $\vee e$ rule to proceed. Basically, we need to arrive at a contradiction when either $P(x_0)$ or $Q(x_0)$ is true. A more detailed proof structure of the internal box is as follows:

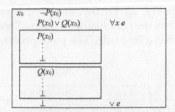

The remaining task is to fill out gaps in each innermost assumption box. It is fairly easy, because we have $\neg P(x_0)$ from $\exists x\ \neg P(x)$ and $\neg Q(x_0)$ from $\forall x\ \neg Q(x)$. Here is the complete proof:

1.	$\forall x\ (P(x) \vee Q(x))$	Premise
2.	$\neg(\forall x\, P(x) \vee \exists x\, Q(x))$	Assumption
3.	$\neg \forall x\, P(x) \wedge \neg \exists x\, Q(x)$	Equivalent, 2
4.	$\neg \forall x\, P(x)$	$\wedge e$, 3
5.	$\neg \exists x\, Q(x)$	$\wedge e$, 3
6.	$\exists x\, \neg P(x)$	Equivalent, 4
7.	$\forall x\, \neg Q(x)$	Equivalent, 5
8.	$x_0 \quad \neg P(x_0)$	Assumption
9.	$P(x_0) \vee Q(x_0)$	$\forall x\ e$, 1
10.	$P(x_0)$	Assumption
11.	$\neg P(x_0)$	Copy, 8
12.	$\bot$	$\bot i$
13.	$P(x_0) \to \bot$	$\to i$, 10–12
14.	$Q(x_0)$	Assumption
15.	$\neg Q(x_0)$	$\forall x\ e$
16.	$\bot$	$\bot i$
17.	$Q(x_0) \to \bot$	$\to i$, 14–16
18.	$\bot$	$\vee e$, 9, 13, 17
17.	$\bot$	$\exists x\ e$, 6, 8–16
18.	$\forall x\, P(x) \vee \exists x\, Q(x)$	PBC, 2–17

The sequents in Examples 5.12 through 5.15 only involve a single variable. The next example shows how to prove a sequent with multiple variables and nested quantifiers.

Example 5.16: Proof of a Sequent with Nested Quantifiers

Show

$$\exists x \exists y\ P(x, y) \vdash \exists y \exists x\ P(x, y)$$

Proof:

By rewriting the premise as $\exists x(\exists y\,P(x, y))$, we know it is in the form $\exists x\varphi$, and thus we can perform $\varphi[x_0/x]$ to start the proof, which is done by opening an assumption box for x_0. Because φ is in the same form, inside the assumption box for x_0 we need to open an internal assumption box for y_0. Under the nested assumptions we should get $P(x_0, y_0)$. Applying $\exists x\ i$ to $P(x_0, y_0)$, we will get $\exists x\,P(x, y_0)$. Further applying $\exists y\ i$ to $\exists x\,P(x, y_0)$, we will get $\exists y\exists x\,P(x, y)$, which is the conclusion to prove. Then we need to exit the boxes one by one with the $\exists x\ e$ rule. The proof process is listed as follows:

1.	$\exists x\exists y\,P(x, y)$			Premise
2.	x_0	$\exists y\,P(x_0, y)$		Assumption
3.		y_0	$P(x_0, y_0)$	Assumption
4.			$\exists x\,P(x, y_0)$	$\exists x\ i,\ 3$
5.			$\exists y\exists x\,P(x, y)$	$\exists x\ i,\ 4$
6.		$\exists y\exists x\,P(x, y)$		$\exists x\ e,\ 2,\ 3\text{--}5$
7.	$\exists y\exists x\,P(x, y)$			$\exists x\ e,\ 1,\ 2\text{--}6$

5.5 Semantics of Predicate Logic

The semantics of predicate logic concern the assignment of meanings to the predicates, variables, constants, and function symbols in the syntax, and the systematical evaluation of the meaning of a formula from the meanings of its constituent parts and the order in which those parts combine.

5.5.1 Interpretation and Models

In the propositional logic, a formula is evaluated by assigning truth values to all its constituent logic atoms first, and then repeatedly applying truth tables for corresponding logic operators. For example, given $p \vee \neg q \rightarrow \neg p \wedge q$, we evaluate it based on four possible truth value assignments to p and q. For each assignment, we apply the truth tables for $\neg$, $\vee$, $\wedge$, and $\rightarrow$ and compute the truth value of the formula. This process is relatively simple. However, for a predicate logic formula, we have much more to do. For example, given $\forall x\,P(x) \vee \exists x\,Q(x)$, we first need to decide the truth values of $\forall x\,P(x)$ and $\exists x\,Q(x)$. We cannot simply assign a truth value to any of them because they are not atomic sentences. Instead, their values depend on the variable x, the predicates P and Q, and specific quantifiers.

An *interpretation* in predicate logic provides semantic meaning to the terms and formulas of the language. Consider the predicate $P(a, f(x))$, where a is a constant, x a variable, f a function symbol, and P a predicate symbol. To find out whether it is true or false, we have to know the meaning of all the symbols in the predicate. The evaluation process proceeds bottom up along the tree shown in Figure 5.3.

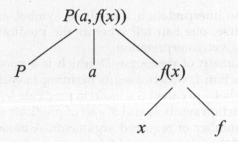

FIGURE 5.3
Evaluation of $P(a, f(x))$.

Variables are objects of predicate logic. In fact, a predicate logic state-ment is about objects. Objects form the domain of discourse D. For example, $\exists x\, P(x)$ states the existence of an object x such that the predicate P is true when referred to it. The domain of discourse is the set of considered objects. For example, if $P(x)$ means "x is a professor," then the domain of discourse here is a set of people, maybe in a specific institute that is indirectly stated in the context.

The interpretation of a function symbol is a function. Each function symbol f of arity n is assigned a function from D^n to D. For example, con-sider a function $f(x, y)$, where the domain of both x and y is the set of inte-gers, the function symbol f can possibly be interpreted as the function that gives the sum of its arguments, the function that finds the larger argu-ment, the function that compares if the two arguments are equal, and so on. Denote by $I(f)$ the interpretation of f. If $f(x, y)$ is to calculate the product of the arguments, then f is associated with multiplication, that is, $I(f) = $ *multiplication*.

A constant symbol is a function symbol with 0 arguments. It is a function that maps a one-element set D^0 to D. A constant can be simply identified with an object in D. For example, if the domain of discourse is a set of peo-ple, an interpretation may assign the value $I(a) = $ "Brown" to the constant symbol a.

The interpretation of an n-ary predicate symbol is a set of n-tuples of ele-ments of the domain of discourse. More precisely, each predicate symbol P of arity n is assigned a relation $I(P)$ over D^n or, equivalently, a function from D^n to $\{true, false\}$. For example, given a domain of discourse

$$D = \{a, b, c, d, e, f, g\}$$

and a binary predicate $P(x, y)$, we may have

$$I(P) = \{(a, b), (a, c), (c, b), (b, d), (e, a), (c, f), (f, g), (b, g), (g, c)\}$$

Therefore, given an interpretation, a predicate symbol, and n objects of the domain of discourse, one can tell whether the predicate is true of those objects under the given interpretation.

Together, the domain of discourse D, which is a nonempty set, and an interpretation function I, which assigns meaning to each function symbol and predicate symbol, are called the *model* in predicate logic. More precisely, let $\mathcal{F}$ be a set of function symbols and $\mathcal{P}$ a set of predicate symbols, each symbol with a fixed number of required arguments. A model $\mathcal{M}$ over the pair of $(\mathcal{F}, \mathcal{P})$ consists of the following set of data:

- A nonempty set D^M, representing the domain of discourse
- A concrete element f^M of D^M for every nullary function symbol (constant) $f \in \mathcal{F}$
- A concrete element $f^M : (D^M)^n \to D^M$ for every n-ary ($n > 0$) function symbol $f \in \mathcal{F}$
- A subset P^M of n-tuples over D^M for every n-ary predicate symbol $P \in \mathcal{P}$

Example 5.17: Interpretation and Model

Let $\mathcal{F} = \{c_f, c_l\}$ and $\mathcal{P} = \{P, M, E\}$, where c_f and c_l are constants, P is a binary predicate, and M and E are both unary predicates. To define a model $\mathcal{M}$, we need to have a data set to serve as the domain of discourse. Let it be a set of major courses in the curriculum of a computer science (CS) program, which is given as

$$D^M = \{\text{CS125, CS175, CS225, CS275, CS325, CS355, CS365, CS375, CS425,}$$
$$\text{CS455, CS465, CS475, CS495}\}$$

As in any academic program, some courses are mandatory, some courses are elective, and some courses are prerequisites to some other courses. The dependencies among courses are shown in Figure 5.4.

The interpretation of the symbols in $\mathcal{F}$ and $\mathcal{P}$ is as follows:

- c_f^M is the first mandatory CS course for students to take.
- c_l^M is the last mandatory CS course for students to take.

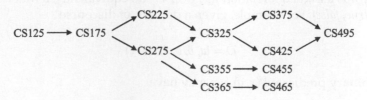

FIGURE 5.4
Course dependencies in Example 5.17.

- P^M specifies the relation of dependency between two courses: the course of first argument is a direct prerequisite for the course of second argument.
- M^M indicates a mandatory course.
- E^M indicates an elective course.

Let

$$c_f^M = CS125,$$
$$c_l^M = CS495,$$
$$P^M = \{(CS125, CS175),$$
$$(CS175, CS225),$$
$$(CS175, CS275),$$
$$(CS225, CS325),$$
$$(CS275, CS325),$$
$$(CS275, CS355),$$
$$(CS275, CS365),$$
$$(CS325, CS375),$$
$$(CS325, CS425),$$
$$(CS355, CS455),$$
$$(CS365, CS465),$$
$$(CS375, CS495),$$
$$(CS425, CS495)\},$$
$$M^M = \{CS125, CS175, CS225, CS275, CS325, CS375, CS425, CS495\},$$
$$E^M = \{CS355, CS365, CS455, CS465\}$$

The relations of P^M is intuitively shown in Figure 5.4. With this model, we can evaluate some formulas:

1. $\forall x\, (M(x) \rightarrow \neg E(x))$
   This formula states that for each course, if it is mandatory, it must not be elective. Examining M^M and E^M, we know the two sets are disjoint (they should be). Therefore, this formula is true in our model.
2. $\exists x\, P(c_f, x)$
   This formula states that there is at least one course that c_f is a prerequisite. Because $c_f^M = CS125$ and $(CS175, CS125) \in P^M$, this formula is true.
3. $\exists x\, (P(x, c_l))$
   This formula is identical to the previous one. It states that there is at least one course that is a prerequisite of c_l. Because $c_l^M = CS495$ and $(CS375, CS495) \in P^M$, this formula is true.
4. $\exists x \forall y\, (\neg P(x, y))$
   This formula states that there is one course that is not a prerequisite for any other courses. Looking at the set of P^M, we know there is no single pair that the first element is CS495. Thus CS495 is the x. The formula is true.
5. $\exists x \exists y \exists z\, (M(y) \wedge E(z) \wedge P(x, y) \wedge P(x, z))$
   This formula states that there is a course that is a prerequisite for a mandatory course and also a prerequisite for an elective course. The way we determine if this is true is simple: if we can

find an entry, say e_1, in the M^M set, a second entry, say e_2, in the E^M set and a third entry, say e_3, in the D^M set, such that $(e_1, e_3) \in P^M$ and $(e_2, e_3) \in P^M$, then this formula is true. A selection of $(e_1 = \text{CS325}, e_2 = \text{CS355}, e_3 = \text{CS275})$ meets the condition, where e_3 is the course represented by x, e_1 the one represented by y, and e_2 the one represented by z, Therefore, the formula is true.

5.5.2 Evaluation of Truth Values

In the previous section, we discussed interpretation of function symbols and predicate symbols, and the model in predicate logic. In Example 5.17, we performed informal evaluation on some predicate logic formulas. We discuss how to evaluate variables and quantifiers now.

Consider the predicate $M(x)$ in the model of Example 5.17. We cannot tell if it is true or false because we don't have a value for x. Obviously, if $x \in M^M$, $M(x)$ is true; if $x \in D^M \backslash M^M$, $M(x)$ is false. We introduce a variable assignment function μ that associates elements of the domain of discourse with each variable. This way,

- $M(x)$ evaluates to true, if $\mu(x) \in M^M$
- $M(x)$ evaluates to false, if $\mu(x) \in D^M \backslash M^M$

One can come up with many different variable assignment functions. For example, for the x in $M(x)$, we can have

$$\mu_1(x) = \text{CS175}, \mu_2(x) = \text{CS375}, \mu_3(x) = \text{CS455}, \dots$$

Under $\mu_1(x)$ and $\mu_2(x)$, $M(x)$ evaluates to true, while under $\mu_3(x)$, $M(x)$ evaluates to false.

Example 5.18: Variable Assignment

Consider the predicate $P(x, y)$ in Example 5.17. Find out which truth value it will evaluate to under each of the following assignment functions:

1. $\mu_1(x) = \text{CS175}, \mu_1(y) = \text{CS225}$
2. $\mu_2(x) = \text{CS275}, \mu_2(y) = \text{CS125}$
3. $\mu_3(x) = \text{CS375}, \mu_1(y) = \text{CS425}$

Solution

1. $P(x, y)$ evaluates to true because $(\text{CS175}, \text{CS225}) \in P^M$.
2. $P(x, y)$ evaluates to false because $(\text{CS275}, \text{CS125}) \notin P^M$.
3. $P(x, y)$ evaluates to true because $(\text{CS375}, \text{CS425}) \notin P^M$.

The variable assignment μ can be extended to all terms, which associates a single element of the domain of discourse to each term.

We have introduced the domain of discourse, function symbol and predicate symbol interpretation, variable assignment, and the model in predicate logic. Now let us formulate the truth evaluation of a general predicate logic formula φ. Assume a model $\mathcal{M}$ and variable assignment function μ. The domain of discourse of the model is D.

1. $\varphi = P(t_1, t_2, ..., t_n)$

   Let $\mu(t_1, t_2, ..., t_n) = (v_1, v_2, ..., v_n)$. φ is true if and only if

   $$(v_1, v_2, ..., v_n) \in P^M$$

   where $v_i \in D^M, i = 1, 2, ..., n$

2. $\varphi = (t_1 = t_2)$

   φ is true if and only if $\mu(t_1) = \mu(t_2)$, where $\mu(t_1) \in D^M$ and $\mu(t_2) \in D^M$

3. $\varphi = (\varphi_1 \wedge \varphi_2)$

   The truth value of φ is determined by the truth values of φ_1 and φ_2 and the application of the conjunction rule in the truth table. This applies to the cases of $(\varphi_1 \vee \varphi_2)$, $(\varphi_1 \rightarrow \varphi_2)$, and $\neg\varphi_1$ as well.

4. $\varphi = \forall x\varphi_1$

   φ is true if and only if φ_1 is true for every possible μ' that differs only in the value of x. This is equal to saying that $\forall x\varphi_1$ is true if and only if every possible choice of the value of x in the domain of discourse makes φ_1 true.

5. $\varphi = \exists x\varphi_1$

   φ is true if and only if there is such a μ' that $\mu'(x)$ makes φ_1 to be true. This is equal to saying that $\exists x\varphi_1$ is true if and only if there is a choice of the value of x in the domain of discourse that makes φ_1 true.

In cases 4 and 5 there is no free x, and thus the truth value of the formula does not depend on the initial variable assignment μ. In all other cases, the truth value of the formula depends on the value assignment of free variables by μ. The above five cases can be applied inductively.

Example 5.19: Evaluation of Truth Values

Consider a predicate system that has a model $\mathcal{M}$ as follows:

$$\mathcal{F} = \{f\}, \text{ where } f \text{ is a binary function}$$

$$\mathcal{P} = \{P, Q, R\}, \text{ where } P \text{ is a unary predicate symbol, } Q \text{ a binary predicate symbol, and } R \text{ a ternary predicate symbol.}$$

$$D^M = N, \text{ the set of natural numbers}$$

$$f^M(x, y) = +(x, y) = x + y$$

$$P^M(x) = \{x \mid x \text{ is even}\}$$

$$Q^M(x, y) = \{(x, y) \mid x > y\}$$

$$R^M(x, y, z) = \{(x, y, z) \mid x + y = z\}$$

The symbols x, y, and z are all variables. The variable assignment function is μ. Evaluate the following formulas.

1. $x = y$
2. $P(x)$
3. $Q(x, f(x, y))$
4. $\forall x \exists y\, Q(x, y)$
5. $\exists x \exists y\, (Q(x, y) \rightarrow Q(y, x))$
6. $\forall x \forall y \forall z\, R(x, y, z)$
7. $\forall x \forall y\, R(x, y, f(x, y))$

Solution

1. In this formula, both x and y are free variables, and thus whether it is true depends on the variable assignment function μ. $x = y$ is true if and only if $\mu(x) = \mu(y)$, that is, μ assigns x and y to the same element in the set D^M.
2. x is free in $P(x)$. If $\mu(x)$ is an even number, then $P(x)$ is true. Otherwise, it is false.
3. Because $f(x, y) = \mu(x) + \mu(y)$ and all members in D^M are greater than 0, regardless of what μ is, it is impossible for $\mu(x) > (\mu(x) + \mu(y))$ to be true. Therefore, $Q(x, f(x, y))$ is false.
4. The formula $\forall x \exists y\, Q(x, y)$ states that for every natural number, there exists another natural number that is greater. It is true, because we can always choose y to as $x + 1$ however we select x.
5. The formula $\exists x \exists y\, (Q(x, y) \rightarrow Q(y, x))$ states that there is an assignment μ' such that

$$\mu'(x) > \mu'(y) \rightarrow \mu'(y) > \mu'(x)$$

   If $(\mu'(x) > \mu'(y))$ is true, we won't have $(\mu'(y) > \mu'(x))$, regardless of what μ' is. However, any μ' that makes $(\mu'(x) > \mu'(y))$ false will make the implication true. We can easily find such a μ'. For example, $\mu'(x, y) = (1, 2)$. Therefore, the formula is true.
6. The formula $\forall x \forall y \forall z\, R(x, y, z)$ states that for any given three nature numbers, the sum of the first two is equal to the third number. It is false because an assignment of $\mu'(x, y, z) = (2, 3, 4)$ would make the formula false.
7. The formula $\forall x \forall y\, R(x, y, f(x, y))$ is true if for any assignment function μ', the following is true:

$$\mu'(x) + \mu'(y) = f^M(\mu'(x), \mu'(x))$$

   Because f^M is defined as the sum of its two arguments, $f^M(\mu'(x), \mu'(x)) = \mu'(x) + \mu'(y)$. Therefore, the formula is true.

5.5.3 Satisfiability and Validity

We mentioned earlier that a well-formed predicate logic formula that contains no free variables is a sentence. Given a model, a sentence evaluates to a fixed truth value. A sentence φ is *satisfiable* if there is a model $\mathcal{M}$ under which it evaluates to true. It this case, we also say $\mathcal{M}$ satisfies φ, and the relation is denoted by $\mathcal{M}|= \varphi$.

Assume P is a binary predicate symbol. Let us check whether or not the sentence

$$\forall x \forall y \, P(x, y)$$

is satisfiable. We answer this question by trying to figure out a model that makes the sentence true. Apparently, the following model $\mathcal{M}$ works:

$$D^{\mathcal{M}} = \{\text{all even numbers}\}$$

$$P^{\mathcal{M}} = \{(x, y) \mid \text{the sum of } x \text{ and } y \text{ is even}\}$$

Therefore, the sentence is satisfiable. Of course, we can come up with many other models that satisfy this sentence.

There are also sentences that are not satisfiable. Those sentences are always false. For example, the two conjuncts in $(\forall x \, R(x)) \wedge (\exists x \neg R(x))$ contradict to each other, and thus the sentence always evaluates to false.

The truth value of a formula with free variables cannot be solely decided based on a model. The assignment of each free variable plays critical role in the truth evaluation, while the free variable assignment function is not part of the model. Consider the simple formula

$$\forall x \, P(x, y)$$

in which y is a free variable. No matter which model is selected, the formula's truth value varies with the element in the domain of discourse assigned to y.

A formula is logically *valid* if it is true in every possible model. An example of such a formula is $\forall x \, R(x) \vee \exists x \neg R(x)$. These formulas play a role similar to tautologies in propositional logic.

Exercises

1. Define predicates

    $S(x)$: *x is a student*

    $P(x)$: *x is a professor*

    $C(x)$: *x is a computer science course*

$T(x, y)$: x teaches y

$L(x, y)$: x learns y

Translate the following statements to predicate logic.

a. Annie is a student.

b. Jim is a professor, so is Bill.

c. Data Structure is a computer science course, so is Operating Systems.

d. Jim teaches some computer science courses.

e. Not all professors teach computer science courses.

f. Not all professors teach all computer science courses.

g. Some students do not learn all computer science courses.

h. Some students learned Data Structure but not Operating Systems.

i. Not all students learn all computer science courses.

j. Some students do not learn any computer science courses.

k. Bill only teaches Data Structure.

l. Jim teaches both Data Structure and Operating Systems.

2. Translate the following English sentences into predicate logic formula. (You will need to define proper predicates, variables, and constants.)

a. Mark is an athlete.

b. John plays basketball.

c. Peter likes Mary.

d. Steve has a dog.

e. All children love chocolate.

f. A whale is a mammal.

g. Barking dogs don't bite.

h. Some barking dogs bite.

i. All birds can fly.

j. Some birds do not fly, but can run.

k. All deer eat grass. Some deer also eat fish.

l. Some cats play with some rats.

m. No animal is both a cat and a dog.

n. Not all students love calculus.

o. A good teacher is always respected by all his students.

    p. Some cars run faster than some other cars.

    q. Some cars run faster than some motorcycles, and all motorcycles run faster than bicycles.

    r. Students take all core courses and some elective courses.

3. Let P be a unary predicate symbol; Q be a 2-ary predicate symbol; R be a 3-ary predicate symbol; x, y, and z be variables; f be a 2-ary function symbol; and m and n be constants. Determine which of the following is a well-formed formula.

    a. $x \rightarrow (x)$

    b. $(x) \wedge Q(x)$

    c. $f(x, y)$

    d. $Q(m, m)$

    e. $((x, y))$

    f. $(x) \rightarrow Q(m, n)$

    g. $f(x, y) \wedge Q(m, n)$

    h. $Q(f(x, y), f(m, n)) \rightarrow R(x, y, z)$

    i. $(f(m, n), n) \vee Q(x, f(m, n))$

    j. $P(x) \rightarrow Q(x, P(y, z))$

    k. $P(x) \wedge Q(x, y) \rightarrow f(m, n)$

    l. $\forall x\, P(x) \vee Q(m, n)$

    m. $\exists x\, Q(x, n)$

    n. $\exists x\, f(x, n)$

    o. $\forall x\, P(x) \vee \exists x\, Q(x, n)$

    p. $\forall x\, P(x) \vee \exists x \forall y\, Q(x, y)$

    q. $P(m) \rightarrow \exists x \forall y\, Q(x, y) \vee \exists y \forall x\, Q(x, y)$

    r. $\forall x\, P(x) \rightarrow \exists x \forall y\, Q(x, y) \rightarrow P(y)$

    s. $\forall x \forall y \exists z\, (R(x, y, z) \wedge \rightarrow Q(x, y))$

    t. $\exists x\, P(x)\, \forall x \exists y\, (Q(x, y)) \rightarrow \forall x \exists y \forall z\, R(x, y, z)$

4. In the following formulas, x, y, and z are variables, and P, Q, and R are predicate symbols. Draw a parse tree for each formula and identify free variables and bound variables.

    a. $P(x) \rightarrow \forall x\, Q(x, y)$

    b. $\forall x \forall y\, (P(x, y) \wedge Q(x) \rightarrow \forall x\, R(x, y))$

    c. $\exists x (P(x) \vee Q(x, y)) \rightarrow \forall x \forall y\, (R(x, y, z) \wedge P(x))$

d. $(x) \land \forall x \forall y\ Q(x, y) \rightarrow (\exists x \forall y\ (P(x) \rightarrow \forall z\ R(x, z)))$

e. $\exists x\ P(x, y) \rightarrow \forall x \forall y\ (Q(x, y, z) \lor \forall z\ R(z))$

5. In the following formulas, x, y, and z are variables; f is a function symbol; m and n are constants; and P, Q, and R are predicate symbols. Draw a parse tree for each formula and identify free variables and bound variables.

a. $\forall x\ P(x, y) \land \exists y\ Q(f(m, n), y)$

b. $(m) \land \forall x\ Q(f(x), y) \rightarrow (\exists x \forall y\ (P(x) \rightarrow \forall z\ R(f(x), z)))$

c. $\forall x \forall y\ (Q(x, y, f(x, y) \land \forall z\ R(z)) \rightarrow \exists x\ P(f(x, f(m, n), y)$

d. $\forall x \forall y\ P(x, y) \lor \exists x\ Q(f(m, n), f(x, f(x, y)))$

e. $\forall x\ (P(x) \lor \forall x \forall y\ Q(f(x), y)) \rightarrow \exists x\ Q(f(y), x)$

6. Find $\varphi_1[c/x]$, $\varphi_2[c/y]$, $\varphi_3[x/y]$, $\varphi_4[x/y]$ and $\varphi_5[(x + c)/y]$, where c is a constant.

a. $\varphi_1 = \forall x \exists y\ (P(x) \land Q(x, y))$

b. $\varphi_2 = \forall x\ (P(x, y)) \rightarrow \exists y\ (Q(y)) \land P(x, y)$

c. $\varphi_3 = P(y) \land \forall y\ Q(x, y) \rightarrow \forall x\ R(x, y)$

d. $\varphi_4 = \forall x \exists y\ (P(x, y) \land Q(x, y)) \rightarrow \exists z\ R(z, y)$

e. $\varphi_5 = P(x, y) \land \forall x \forall y\ Q(x, y) \land \exists x\ R(z, y)$

7. Prove the validity of the following sequents in predicate logic.

a. $\forall x\ (P(x) \rightarrow Q(x)),\ \forall x\ P(x) \vdash \forall x\ Q(x)$

b. $\forall x\ (P(x) \lor Q(x)),\ \forall x\ \neg P(x) \vdash \forall x\ Q(x)$

c. $\exists x\ (P(x) \lor Q(x)),\ \forall x\ \neg P(x) \vdash \exists x\ Q(x)$

d. $\forall x\ (P(x) \land Q(x)) \vdash \forall x(\neg P(x) \rightarrow Q(x))$

e. $\forall x\ (P(x) \land Q(x)),\ \forall x\ R(x) \vdash \forall x\ (P(x) \land Q(x) \land R(x))$

f. $\forall x\ (P(x) \land Q(x)),\ \exists x\ R(x) \vdash \exists x\ (P(x) \land Q(x) \land R(x))$

g. $\exists x\ (P(x) \rightarrow Q(x)),\ \forall x\ P(x) \vdash \exists x\ Q(x)$

h. $\exists x\ (P(x) \rightarrow Q(x)) \vdash \exists x\ (\neg P(x) \lor Q(x))$

8. Prove the validity of the following sequents in predicate logic.

a. $\exists x\ \neg P(x) \vdash \neg \forall x\ P(x)$

b. $\forall x\ \neg P(x) \vdash \neg \exists x\ P(x)$

c. $\exists x\ P(x) \lor \exists x\ Q(x) \vdash \exists x\ (P(x) \lor Q(x))$

d. $\forall x\ P(x) \lor \forall x\ Q(x) \vdash \forall x\ (P(x) \lor Q(x))$

9. Let P be a unary predicate symbol. Find a model that satisfies $\forall x\ P(x)$; also find a mode that does not.

10. Let P be a unary predicate symbol and Q a binary predicate symbol. Find a model that satisfies $\forall x(P(x) \rightarrow \exists y\ Q(x, y))$; also find a mode that does not.

11. Consider a predicate system that has a model $\mathcal{M}$ as follows:

$\mathcal{F} = \{f, g\}$, where f and g are binary functions,

$\mathcal{P} = \{P, Q\}$, where P is a unary predicate symbol and Q a binary predicate symbol.

$$D^M = N, \text{ the set of odd numbers}$$

$$f^M(x, y) = -(x, y) = x - y$$

$$g^M(x, y) = {}^*(x, y) = x * y$$

$$P^M(x) = \{x \,|\, x \text{ is even}\}$$

$$Q^M(x, y) = \{(x, y) \,|\, x < y\}$$

x and y are variables. The variable assignment function is $\mu(x, y) = (5, 7)$. Evaluate the following formulas.

a. $P(f(x, y))$
b. $P(g(x, y))$
c. $Q(x, f(x, y))$
d. $\forall x \exists y \, Q(x, y)$
e. $\forall x \forall y \, Q(f(x, y), g(x, y))$
f. $\exists x \exists y \, (Q(x, y) \wedge Q(g(x, y), y))$

6

Temporal Logic

Propositional logic assigns a truth value to a logic variable. When evaluating a logic formula, we assume each variable has a fixed truth value. By contrast, temporal logic assigns a truth value to a variable at each point in time. In other words, a single variable in temporal logic may have different logic values at different points of time. Temporal logic is a formal system for specifying and reasoning about a system's dynamic properties. This chapter introduces three types of temporal logic: linear temporal logic (LTL), computation tree logic (CTL), and a superset of LTL and CTL called CTL*. Historically, LTL was first proposed for the verification of computer programs by Amir Pnueli in 1977. Four years later in 1981, E. M. Clarke and E. A. Emerson invented CTL and CTL model checking. CTL* was defined by E. A. Emerson and Joseph Y. Halpern in 1986.

6.1 Temporal Logic

Temporal logic is a branch of formal logic that consists of rules and symbolism for representing and reasoning about propositions qualified in terms of time. It describes the ordering of events in time without introducing time explicitly. A temporal logic model contains temporal states. A logic variable or formula can be true in one state but false in another state. The set of states correspond to moments in time. How we navigate between these states depends on our particular view of time.

6.1.1 Kripke Structures

Temporal logic is traditionally interpreted in terms of *Kripke structures*. A Kripke structure is a kind of state transition system that represents the behavior of a system. Assume a fixed set Σ of atomic propositions. For a set S of states, let L be a *labeling function* that maps S to the power set of Σ. The power set of Σ, denoted by 2^Σ, is the set of all subsets of Σ. For example, if $\Sigma = \{p, q, r\}$, then

$$2^\Sigma = \left\{\varnothing, p, q, r, \{p,q\}, \{q,r\}, \{p,r\}, \{p,q,r\}\right\}$$

For each individual state s, $L(s)$ is a set of all atomic propositions that are evaluated to true in the state. Mathematically, a Kripke structure is defined as $M = (S, I, R, L)$, where

- S is a set of states.
- I is a set of initial states. $I \subseteq S$.
- $R \subseteq S \times S$ is a set of state transition relations. For each $s \in S$, there is s' such that $s \to s'$ and the relation is denoted as $(s, s') \in R$.
- L is a labeling function that maps S to the power set of Σ, where Σ is the set of atomic propositions.

The transition system of a Kripke structure can be more intuitively illustrated as a directed graph, in which each node is a state, state transitions are depicted by directed arrows, and the labeling of each state is marked on the node.

Example 6.1: Kripke Structure

In the transition system shown in Figure 6.1, $\Sigma = \{p, q, r\}$, and the Kripke structure is

$$S = \{s_1, s_2, s_3, s_4\}$$

$$I = \{s_1\}$$

$$R = \{(s_1, s_2), (s_1, s_3), (s_2, s_1), (s_2, s_4), (s_3, s_4), (s_4, s_3)\}$$

$$L(s_1) = \{p, r\}$$

$$L(s_2) = \{p, q\}$$

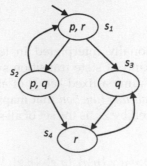

FIGURE 6.1
Directed graph of a transition system.

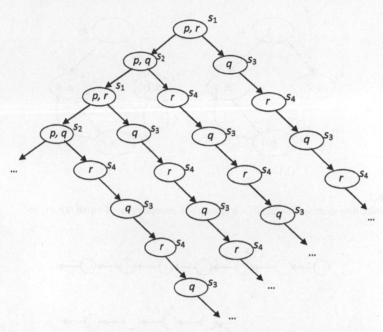

FIGURE 6.2
Computation paths of the system shown in Figure 6.1.

$$L(s_3) = \{q\}$$

$$L(s_4) = \{r\}$$

Unwinding the transition system results in an infinite tree of all possible computation paths, as shown in Figure 6.2.

Notice that the definition of R in a Kripke structure indicates that each computation path is infinite. If a state that represents a deadlock situation exists in a transition system, we can draw an outgoing edge back to the state. For example, for the transition system shown in Figure 6.3a, the state s_2 represents a deadlock. Figure 6.3b shows the system in Kripke structure, in which a self-loop on s_2 is introduced.

6.1.2 Modeling of Time

There are two models of time in temporal logic. One model thinks time as a path of time points. For any two distinguished points along the path, one must be earlier than the other. Mathematically, we represent time as a structure $(T, <)$, where T is a set of time points and $<$ is a precedence relation on T. If a pair (s, t) belongs to $<$, we say that s is earlier than t, which is denoted as $s < t$. For a point s, the set $\{t \in T | s < t\}$ is called the *future* of s; the past

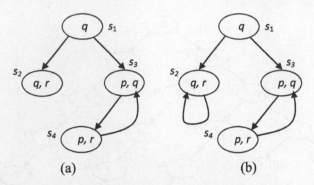

FIGURE 6.3
(a) A transition system with a deadlock state. (b) Corresponding Kripke structure.

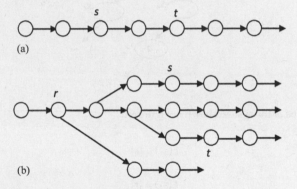

FIGURE 6.4
Modeling of time: (a) Linear time. (b) Branch time.

of s is defined likewise. This model is called *linear-time model*, because all time points are linearly ordered. The linear-time model is illustrated in Figure 6.4a.

The second model is based on a *branching-time* structure. In this structure, a time point, say r, may have two or more future time points that are *not* related to each other. In other words, for any two of these future points, say s and t, we cannot say that s is the future of t or vice versa. This also means that the future of r is not deterministic; or r branches out to the future. The branching-time structure is of a tree structure, rooted at the present moment, which is illustrated in Figure 6.4b.

Linear-time temporal logic and branching-time temporal logic correspond to two distinct views of time. As a consequence, they are expressively incomparable. We will discuss this further after we have introduced linear temporal logic and computation tree logic.

6.2 Linear Temporal Logic

Linear temporal logic (LTL) models time as a sequence of time points that extends infinitely into the future. A time point is represented by a state. A sequence of states is also called a *computation path*, or simply a *path*. Future is not determined, so there is more than one path, representing different possible futures.

6.2.1 Syntax of LTL

LTL is an extension to classical propositional logic. It extends propositional logic by introducing a set of temporal operators that navigate between states. The temporal operators in LTL include **X, F, G, U** and **R**, where

1. **X** means neXt state. **X** p is true if p is true in the next state.
2. **F** means some Future state. **F** p is true if there is a reachable future state in which p is true.
3. **G** means all future states (globally). **G** p is true if p is true in all future states.
4. **U** means until. p **U** q is true if p is true until q is true in a future state.
5. **R** means release. p **R** q is true if q is true until the first state in which p is true.

Assume a fixed set $\sum$ of atomic propositions and let $p \in \sum$. An LTL formula can be inductively defined in BNF (Backus-Naur form) as

$$\varphi ::= \top \mid \bot \mid p \mid (\neg\varphi) \mid (\varphi \wedge \varphi) \mid (\varphi \vee \varphi) \mid (\varphi \rightarrow \varphi)$$
$$\mid (\mathbf{X}\varphi) \mid (\mathbf{F}\varphi) \mid (\mathbf{G}\varphi) \mid (\varphi\mathbf{U}\varphi) \mid (\varphi\mathbf{R}\varphi) \tag{6.1}$$

where $\top$ stands for *true* and $\bot$ stands for *false*. Equation 6.1 states that

1. The two logic constants *true* and *false* are LTL formulas.
2. Any atomic propositional formula is an LTL formula.
3. If φ and ψ are LTL formulas, then so are $\neg\varphi$, $\varphi \wedge \psi$, $\varphi \vee \psi$, and $\varphi \rightarrow \psi$.
4. If φ and ψ are LTL formulas, then so are $\mathbf{X}\varphi$, $\mathbf{F}\varphi$, $\mathbf{G}\varphi$, φ **U** ψ, and φ **R** ψ.

The LTL definition also indicates that $\neg$, **X**, **F**, and **G** are unary operators, while all others are binary operators. Unary operators bind stronger than binary operators do. All the unary operators bind equally strongly. The temporal binary operators **U** and **R** bind more strongly than the propositional logic operators $\wedge$ and $\vee$. The implication operator $\rightarrow$ has the least binding priority. For example, the following two formulas are equal:

$$\neg p \vee (\mathbf{G}\, q \rightarrow \neg r)\, \mathbf{U}\, (p \wedge r)$$

$$(\neg p) \vee (((\mathbf{G}\, q) \rightarrow (\neg r))\, \mathbf{U}\, (p \vee r))$$

An LTL formula is *syntactically correct* or *well-formed* if and only if it obeys the inductive construction rule given in the definition (Equation 6.1). For example, the following formulas are well-formed:

1. $q \rightarrow \neg(p \vee \mathbf{G}\, q) \wedge \mathbf{F}\, p$
2. $\mathbf{G}(p \rightarrow \mathbf{F}\, r) \rightarrow \neg(p \vee q)$
3. $\mathbf{X}\, (q\, \mathbf{R}\, r)$

The following formulas, however, are not well-formed:

1. $\mathbf{F}\, p \rightarrow \mathbf{U}\, r$
2. $p\, \mathbf{X}\, r \rightarrow \mathbf{G}\, q$
3. $p\, \mathbf{U}\, (\wedge\, r)$

In the first formula, the mistake is that the operator $\mathbf{U}$ is used as a unary operator. In the second formula, the next operator $\mathbf{X}$ is used as a binary one. In the last one, the conjunction operator has only a right operand.

6.2.2 Parse Trees of LTL Formulas

A parse tree of an LTL formula is a nested list, where each branch is either a single atomic proposition, or a formula composed of either two propositions and a binary operator, or one proposition and a unary operator. Parse trees are helpful for the truth value evaluation of LTL formulas. The root of the parse tree of an LTL formula is the operator that should be evaluated last, while all leaf nodes of the tree are atomic propositions. If the formula only has an atomic proposition, then the proposition is the root and only node. Parse trees are evaluated from the bottom up.

Example 6.2: Parse Tree

Consider the following formula:

$$\neg p \wedge (\mathbf{G}\, q \rightarrow \neg r)\, \mathbf{U}\, (p \vee r)$$

In this formula, the only $\wedge$ operator has the lowest binding priority and is evaluated last. Thus it is the root of the parse tree. Its left branch is the

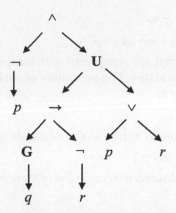

FIGURE 6.5

Parse tree of the formula $\neg p \wedge (\mathbf{G}\, q \rightarrow \neg r)\, \mathbf{U}\, (p \vee r)$.

subtree of $\neg p$ and its right branch is the subtree of $(\mathbf{G}\, q \rightarrow \neg r)\, \mathbf{U}\, (p \vee r)$. The operator $\mathbf{U}$ is the root of the right subtree, with $(\mathbf{G}\, q \rightarrow \neg r)$ and $(p \vee r)$ being its two subbranches. The left subbranch of $(\mathbf{G}\, q \rightarrow \neg r)$ is further a tree with $\rightarrow$ being its root, and $(\mathbf{G}\, q)$ and $(\neg r)$ being its two branches. Continuing the parsing recursively until all operators are hit, we end up with a parse tree depicted in Figure 6.5.

6.2.3 Semantics of LTL

LTL formulas are evaluated over computation paths. Consider a computation path

$$\pi = s_1 \rightarrow s_2 \rightarrow \cdots$$

which starts from the state s_1. Denote by π_i the path starting with s_i, that is,

$$\pi_i = s_i \rightarrow s_{i+1} \rightarrow \cdots$$

We now define the binary *satisfaction relation*, denoted by $\vDash$, for LTL formulas. The satisfaction is with respect to a computation path π. We always have

- $\pi \vDash \top$
- $\pi \nvDash \bot$

For any single atomic proposition $p \in \Sigma$, we have $\pi \vDash p$ if and only if ("iff" for short) $p \in L(s_1)$. For any LTL formula φ, we have $\pi \vDash \varphi$ iff φ evaluates to true in s_1. Any composition of φ and ψ with propositional logic operators ($\neg$, $\wedge$, $\vee$, and $\rightarrow$) is evaluated in the first state s_1 of the path. Specifically,

1. $\pi \vDash \neg \varphi$ iff $\pi \nvDash \varphi$
2. $\pi \vDash \varphi \wedge \psi$ iff $\pi \vDash \varphi$ and $\pi \vDash \psi$

3. $\pi \vDash \varphi \vee \psi$ iff $\pi \vDash \varphi$ or $\pi \vDash \psi$

4. $\pi \vDash \varphi \rightarrow \psi$ iff $\pi \vDash \psi$ as long as $\pi \vDash \varphi$

   All LTL formulas that are composed with temporal operators should be evaluated across states. The semantics of each temporal operation is defined as follows.

5. $\pi \vDash \mathbf{X}\varphi$ iff $\pi_2 \vDash \varphi$

   That is, iff φ is evaluated to true in the next state s_2, then we have $\pi \vDash \mathbf{X}\varphi$.

6. $\pi \vDash \mathbf{G}\varphi$ iff $\pi_i \vDash \varphi$ for all $i \geq 1$

   That is, iff φ is evaluated to true in every state along the path π, then we have $\pi \vDash \mathbf{G}\varphi$.

7. $\pi \vDash \mathbf{F}\varphi$ iff $\pi_i \vDash \varphi$ for some $i \geq 1$

   That is, iff φ is evaluated to true in some state along the path π, then we have $\pi \vDash \mathbf{F}\varphi$.

8. $\pi \vDash \varphi \mathbf{U}\psi$ iff $\pi_i \vDash \psi$ for some $i \geq 1$ and $\pi_j \vDash \varphi$ for all $j = 1, 2, \ldots i\text{-}1$

   That is, iff φ is evaluated to true in every state along the path π until a state in which ψ is evaluate to true, then we have $\pi \vDash \varphi \mathbf{U}\psi$. By the way, we can view $\mathbf{F}\varphi$ as an abbreviation of $\top \mathbf{U}\varphi$.

9. $\pi \vDash \varphi \mathbf{R}\psi$ iff $\pi_i \vDash \varphi$ for some $i \geq 1$ and $\pi_j \vDash \psi$ for all $j = 1, 2, \ldots i$

   That is, iff ψ is evaluated to true under every state along the path π until a state under which both φ and ψ are evaluate to true, then we have $\pi \vDash \varphi \mathbf{R}\psi$.

Figure 6.6 illustrates the semantics of all these temporal operators. The first bubble in each path represents the first state of the path.

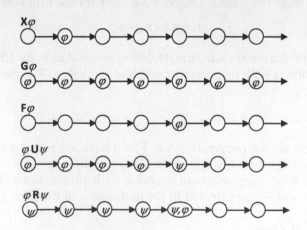

FIGURE 6.6
Illustration of the semantics of LTL temporal operators.

As we can see from Example 6.1, a system may have many or even an infinite number of computation paths. When we verify a system against an LTL formula from a state (typically an initial state), we check if the formula is satisfied by *all* paths from the state. We say

$$M, s \vDash \varphi$$

iff $\pi \vDash \varphi$ holds for every computation path π starting at s.

Example 6.3: LTL Formula Verification

Consider the Kripke structure shown in Figure 6.1. We want to verify $M, s_1 \vDash \varphi$ for a few properties φ specified with the following LTL formulas:

1. $\varphi = p \wedge r$

   Because φ is essentially a propositional formula, $M, s_1 \vDash \varphi$ is evaluated only in s_1. Because $L(s_1) = \{p, r\}$, so $M, s_1 \vDash p \wedge r$ holds.

2. $\varphi = p \wedge q$

   Because s_1 does not satisfy q, $M, s_1 \vDash p \wedge q$ does not hold.

3. $\varphi = \mathbf{X} \ q$

   We check to see if along every path the next state has q. The initial state s_1 has two next states, namely s_2 and s_3. Because both s_2 and s_3 have q, $M, s_1 \vDash \mathbf{X} \ q$ holds.

4. $\varphi = \mathbf{F} \ r$

   We check to see if r is true in some state along every path. From Figure 6.1 we know this is the case. Therefore, $M, s_1 \vDash \mathbf{F} \ r$ holds.

5. $\varphi = \mathbf{G} \ (q \vee r)$

   In the four states, s_1 and s_4 have r, while s_2 and s_3 have q. Therefore, $M, s_1 \vDash \mathbf{G} \ (q \vee r)$ holds.

6. $\varphi = \mathbf{G}(\neg p \rightarrow q \vee r)$

   We check to see if for each state that satisfies $\neg p$, it also satisfies $q \vee r$. s_3 and s_4 satisfies $\neg p$, and s_3 satisfies q while s_4 satisfies r. Therefore, $M, s_1 \vDash \mathbf{G}(\neg p \rightarrow q \vee r)$ holds.

7. $\varphi = \mathbf{FG} \ r$

   We check to see if there is a state in every path such that starting that state $\mathbf{G} \ r$ is true. Figure 6.2 shows that there is no single path that satisfies $\mathbf{G} \ r$ from some point along the path. Therefore, $M, s_1 \vDash \mathbf{FG} \ r$ does not hold.

8. $\varphi = \mathbf{GF} \ r$

   We check to see if $\mathbf{F} \ r$ is true in every state along every path, or in other words, if r is satisfied *infinitely often*. We know this is true because either s_1 or s_4 appears in each path infinitely often. Therefore, $M, s_1 \vDash \mathbf{GF} \ r$ holds.

9. $\varphi = p \ \mathbf{U} \ q$

   For all paths, p is true in the first state and q is true in the second state, which satisfies $p \ \mathbf{U} \ q$. Therefore, $M, s_1 \vDash p \ \mathbf{U} \ q$ holds.

10. $\varphi = (p \wedge r) \ \mathbf{U} \ (p \wedge q)$

Figure 6.1 shows that $p \wedge q$ is true only in some states along the leftmost path. In all other paths it has never been true. Taking the rightmost path, however, $p \wedge r$ is true only in the first state. In the second state, neither $p \wedge r$ nor $p \wedge q$ is true. Therefore, $M, s_1 \vDash (p \wedge r) \ \mathbf{U} \ (p \wedge q)$ does not hold.

11. $\varphi = \mathbf{F} \ q \rightarrow \mathbf{F} \ (p \wedge r)$

This formula states that if any path starting from s_1 satisfies $\mathbf{F} \ q$, the path also satisfies $\mathbf{F} \ (p \wedge r)$. Figure 6.2 shows every path satisfies $\mathbf{F} \ q$, but only the leftmost path satisfies $\mathbf{F} \ (p \wedge r)$. Therefore, $\varphi = \mathbf{F} \ q \rightarrow \mathbf{F} \ (p \wedge r)$ does not hold.

12. $\varphi = \mathbf{GF} \ q \rightarrow \mathbf{GF} \ r$

This property states that if any path starting from s_1 satisfies $\mathbf{GF} \ q$, the path also satisfies $\mathbf{GF} \ r$. The directed graph shows that every single path starting from s_1 satisfies $\mathbf{GF} \ q$; meanwhile, each such a path also satisfies $\mathbf{GF} \ r$. Therefore, $M, s_1 \vDash \mathbf{GF} \ q \rightarrow \mathbf{GF} \ r$ holds.

6.2.4 Equivalencies of LTL Formulas

Two LTL formulas φ and ψ are said to be *semantically equivalent*, denoted by $\varphi \equiv \psi$, if for all models M and all states s in M,

$$M, s \vDash \varphi \ iff \ M, s \vDash \psi$$

Simply put, two formulas are equivalent if they are evaluated to the same truth value from any state of any computation path of any Kripke structure. Listed below are a few equivalencies:

$$\mathbf{X}(\varphi \wedge \psi) \equiv \mathbf{X}\varphi \wedge \mathbf{X}\psi$$

$$\mathbf{X}(\varphi \vee \psi) \equiv \mathbf{X}\varphi \vee \mathbf{X}\psi$$

$$\mathbf{X}(\varphi \ \mathbf{U} \ \psi) \equiv \mathbf{X}\varphi \ \mathbf{U} \ \mathbf{X}\psi$$

$$\mathbf{F}(\varphi \vee \psi) \equiv \mathbf{F}\varphi \vee \mathbf{F}\psi$$

$$\mathbf{G}(\varphi \wedge \psi) \equiv \mathbf{G}\varphi \wedge \mathbf{G}\psi$$

$$\neg\mathbf{X}\varphi \equiv \mathbf{X}\neg\varphi$$

$$\mathbf{F} \ \mathbf{F}\varphi \equiv \mathbf{F}\varphi$$

$$\mathbf{G} \ \mathbf{G}\varphi \equiv \mathbf{G}\varphi$$

$$\neg\mathbf{F}\varphi \equiv \mathbf{G}\neg\varphi$$

$$\mathbf{F}\varphi \equiv \top \ \mathbf{U} \ \varphi$$

$$\neg(\varphi \ \mathbf{R} \ \psi) \equiv \neg\varphi \ \mathbf{U} \ \neg\psi$$

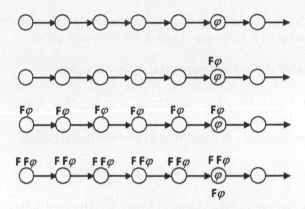

FIGURE 6.7
Illustration of $\mathbf{F}\,\mathbf{F}\varphi \equiv \mathbf{F}\varphi$.

The last three equivalencies indicate that $\mathbf{F}$ and $\mathbf{G}$ are a dual; $\mathbf{F}$ and $\mathbf{U}$ are a dual; and $\mathbf{R}$ and $\mathbf{U}$ are a dual. Therefore, any LTL formula can be converted with one that only contains $\mathbf{X}$, $\mathbf{F}$, and $\mathbf{U}$, the three temporal operators plus all Boolean operators.

All these equivalencies result from the semantics of LTL operators. Figure 6.7 illustrates $\mathbf{F}\,\mathbf{F}\varphi \equiv \mathbf{F}\varphi$. Assume a state in a path satisfies φ (the first diagram). According to the definition of $\mathbf{F}$, the state also satisfies $\mathbf{F}\varphi$. Therefore, we can mark the state with $\mathbf{F}\varphi$ (the second diagram). Again according to the definition of $\mathbf{F}$, we can further mark all previous states along the path with $\mathbf{F}\varphi$ (the third diagram). By the same token, since the state satisfies $\mathbf{F}\varphi$, we can mark the state and all its previous states with $\mathbf{F}\mathbf{F}\varphi$ (the fourth path). This proves the equivalence.

6.2.5 System Property Specification

We use an elevator of five floors as an example to show how we can use LTL formulas to code the properties of real-world systems. First we define the following atomic propositions:

- *door_closed*: the door of the elevator is closed. Accordingly, ¬*door_closed* means the door is open.
- *direction_up*: the elevator is moving upward.
- *direction_down*: the elevator is moving downward. Accordingly, *direction_up* ∨ *direction_down* means that the elevator is moving.
- *floor_[n]*: the elevator is on the *n*-th floor. For example, *floor_3* means that it is on the 3rd floor.
- *button_[n]*: the *n*-th floor is requested.

With these atomic propositions, we can formally specify some properties of the elevator using LTL formulas. Here are a few examples:

1. The elevator should not move if the door is open.

$$\mathbf{G}(\neg door\_closed \rightarrow \neg(direction\_up \vee direction\_down))$$

2. Whenever the door is open, it will eventually be closed.

$$\mathbf{G}(\neg door\_closed \rightarrow \mathbf{F}\, door\_closed)$$

3. Likewise, whenever the door is closed, it will eventually be open.

$$\mathbf{G}(door\_closed \rightarrow \mathbf{F}\, \neg door\_closed)$$

4. The elevator can move upward only if the floor is not the highest.

$$\mathbf{G}(direction\_up \rightarrow \neg floor\_5)$$

5. Similarly, the elevator can move downward only if the floor is not the lowest.

$$\mathbf{G}(direction\_down \rightarrow \neg floor\_1)$$

6. The elevator can visit any floor infinitely often. For example,

$$\mathbf{GF}\, floor\_4$$

7. When a floor is requested, the elevator will eventually stop at the floor. For example:

$$\mathbf{G}(button\_3 \rightarrow \mathbf{F}\, floor\_3)$$

8. When the elevator is traveling upward, it does not change its direction when there are passengers waiting to go to a higher floor. For example:

$$\mathbf{G}\,((floor\_1 \vee floor\_2 \vee floor\_3) \wedge direction\_up \wedge button\_4$$

$$\rightarrow direction\_up\ \mathbf{U}\, floor\_4)$$

9. Likewise, when the elevator is traveling downward, it does not change its direction when it has passengers waiting to go to a lower floor. For example:

$$\mathbf{G}\,((\mathit{floor\_4} \vee \mathit{floor\_5} \vee \mathit{floor\_3}) \wedge \mathit{direction\_down} \wedge \mathit{button\_2}$$

$$\rightarrow \mathit{direction\_down}\ \mathbf{U}\ \mathit{floor\_2})$$

6.3 Computation Tree Logic

LTL formulas are evaluated on paths. We say a state of a system satisfies an LTL formula if *all paths* from the state satisfy it. If we want to specify that there exists a path that satisfies some property φ, it cannot be done directly with LTL. One way to work around this is verify if all paths satisfy $\neg\varphi$. A positive answer to the new problem is a negative answer to the original problem, and vice versa.

Branching-time logics solve this problem with the capability of quantifying explicitly over paths. Computation tree logic (CTL) is a branching-time logic; it models time as a treelike structure in which the future is not determined.

6.3.1 Syntax of CTL

CTL extends traditional propositional logic with extra CTL temporal operators. As shown in Figure 6.8, each CTL temporal operator is a pair of symbols. The first one is a *path quantifier*. It can be either **A** (for *"all* paths") or **E** ("there *exists* a path"). The second one is a linear-time operator and can be **X** ("next state"), **F** ("in a *future* state"), **G** ("*globally* in the future"), or **U** (for *until*), exactly as defined in LTL.

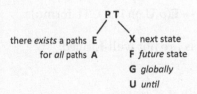

FIGURE 6.8
CTL operators. P: path quantifiers; T: linear-time operators.

Let $\sum$ be a set of atomic propositional formulas and $p \in \sum$. A CTL formula is defined inductively in BNF as

$$\varphi ::= \top \sqcup \mid p \mid (\neg\varphi) \mid (\varphi \wedge \varphi) \mid (\varphi \vee \varphi) \mid (\varphi \to \varphi)$$

$$\mid (\mathbf{AX}\varphi) \mid (\mathbf{EX}\varphi) \mid (\mathbf{AG}\varphi) \mid (\mathbf{EG}\varphi) \mid (\mathbf{AF}\varphi) \mid (\mathbf{EF}\varphi) \qquad (6.2)$$

$$\mid \mathbf{A}(\varphi \mathbf{U} \varphi) \mid \mathbf{E}(\varphi \mathbf{U} \varphi)$$

The first row of the expression states that the two logic constants *true* and *false*, any atomic propositions, and any compound propositional formulas are CTL formulas. The rest of the expression states that if φ and ψ are two CTL formulas, then so are $\mathbf{AX}\varphi$, $\mathbf{EX}\varphi$, $\mathbf{AG}\varphi$, $\mathbf{EG}\varphi$, $\mathbf{AF}\varphi$, $\mathbf{EF}\varphi$, $\mathbf{A}(\varphi \mathbf{U} \psi)$, and $\mathbf{E}(\varphi \mathbf{U} \psi)$.

In CTL formulas, the unary operators $\neg$, $\mathbf{AG}$, $\mathbf{EG}$, $\mathbf{AF}$, $\mathbf{EF}$, $\mathbf{AX}$, and $\mathbf{EX}$ have the highest binding priority, then $\wedge$ and $\vee$, then $\mathbf{AU}$ and $\mathbf{EU}$, and $\to$ comes last. The following formulas are well-formed CTL formulas:

1. $\mathbf{AF}(p \to \mathbf{EG}\ (p \vee r))$
2. $\mathbf{E}((p \wedge q)\ \mathbf{U}\ (\mathbf{AG}\ r))$
3. $\mathbf{EF}\ \mathbf{A}(p\ \mathbf{U}\ q)$
4. $\mathbf{AF}\ (\mathbf{EG}\ (q \to r) \to \mathbf{EF}\ p)$
5. $\mathbf{EF}\ (\neg p \wedge r) \to \mathbf{EF}(q \to \mathbf{E}(p\ \mathbf{U}\ q))$

Let us show how Formula 5 is constructed:

p, r, q are CTL formulas
$\neg p$ is a CTL formula
$\neg p \wedge r$ is a CTL formula
$\mathbf{EF}\ (\neg p \wedge r)$, $\mathbf{E}(p\ \mathbf{U}\ q)$ are CTL formulas
$q \to \mathbf{E}(p\ \mathbf{U}\ q)$ is a CTL formula
$\mathbf{EF}(q \to \mathbf{E}(p\ \mathbf{U}\ q))$ is a CTL formula
$\mathbf{EF}\ (\neg p \wedge r)\ \to \mathbf{EF}(q \to \mathbf{E}(p\ \mathbf{U}\ q))$ is a CTL formula

The following formulas are not well-formed:

1. $\mathbf{AF}(p \wedge q \to \mathbf{G}\ r)$
2. $(p \vee q)\ \mathbf{U}\ (\mathbf{EF}\ r)$
3. $\mathbf{EF}\ (p\ \mathbf{U}\ q)$
4. $\mathbf{AG}\ (\mathbf{EF}\ p \to \mathbf{A}(q \to r))$
5. $\neg p \wedge r \to \mathbf{EF}(q \to \mathbf{E}(p \to q))$

In formula 1, the linear-time operator **G** does not have a prefix path quantifier. In formulas 2 and 3, the linear-time operator **U** does not have a prefix path quantifier. In formulas 4 and 5, the path quantifier **A** and **E** do not have successor linear-time operators, respectively.

We can construct a parse tree for any well-formed CTL formula. In a CTL formula parse tree, all leaf nodes are atomic propositions, while all internal nodes are CTL operators. Notice that although we write $\mathbf{A}(\varphi\ \mathbf{U}\ \psi)$ and $\mathbf{E}(\varphi\ \mathbf{U}\ \psi)$ for **AU** and **EU** operations, respectively, as a convention, we draw **AU** or **EU** in a single node and draw φ and ψ as its left child node and right child node, respectively, in the parse tree. Figure 6.9 shows the parse tree of $\mathbf{A}((\neg p \wedge r)\ \mathbf{U}\ \mathbf{EG}(q \rightarrow \mathbf{E}(p\ \mathbf{U}\ q)))$.

6.3.2 Semantics of CTL

Like LTL, the semantics of CTL is defined in terms of states. CTL formulas are evaluated over Kripke structures. Let $M = (S, I, R, L)$ be a Kripke structure, $s \in S$, and φ and ψ be CTL formulas. The satisfaction relation $M, s \models \varphi$ is defined as follows:

1. $M, s \models \top$ and $M, s \not\models \bot$
2. $M, s \models p$ iff $p \in L(s)$
3. $M, s \models \neg\varphi$ iff $M, s \not\models \varphi$
4. $M, s \models \varphi \wedge \psi$ iff $M, s \models \varphi$ and $M, s \models \psi$
5. $M, s \models \varphi \vee \psi$ iff $M, s \models \varphi$ or $M, s \models \psi$
6. $M, s \models \varphi \rightarrow \psi$ iff $M, s \models \psi$ whenever $M, s \models \varphi$

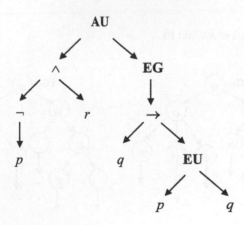

FIGURE 6.9

Parse tree of the formula $\mathbf{A}(\neg p \wedge r\ \mathbf{U}\ \mathbf{EG}(q \rightarrow \mathbf{E}(p\ \mathbf{U}\ q)))$.

7. $M,s \vDash \mathbf{AX} \ \varphi$ iff $M,s' \vDash \varphi$ for all s' such that $(s,s') \in R$

8. $M,s \vDash \mathbf{EX} \ \varphi$ iff $M,s' \vDash \varphi$ for some s' such that $(s,s') \in R$

9. $M,s \vDash \mathbf{AG} \ \varphi$ iff $M,s_i \vDash \varphi$ for any state s_i along any path

10. $M,s \vDash \mathbf{EG} \ \varphi$ iff $M,s_i \vDash \varphi$ for some path and for any state s_i along the path

11. $M,s \vDash \mathbf{AF} \ \varphi$ iff there is a state s_i along every path such that $M,s_i \vDash \varphi$

12. $M,s \vDash \mathbf{EF} \ \varphi$ iff there is a path and for some state s_i along the path such that $M,s_i \vDash \varphi$

13. $M,s \vDash \mathbf{A}(\varphi \ \mathbf{U} \ \psi)$ iff for all paths $\varphi \ \mathbf{U} \ \psi$ is satisfied

14. $M,s \vDash \mathbf{E}(\varphi \ \mathbf{U} \ \psi)$ there is a path in which $\varphi \ \mathbf{U} \ \psi$ is satisfied

The semantics of **AX** and **EX**, **AG** and **EG**, **AF** and **EF**, and **AU** and **EU** are intuitively illustrated in Figures 6.10 through 6.13. Figure 6.12 also shows that **AF** φ is an abbreviation of $\mathbf{A}(\top \ \mathbf{U} \ \varphi)$ and **EF**φ an abbreviation of $\mathbf{E}(\top \ \mathbf{U} \ \varphi)$.

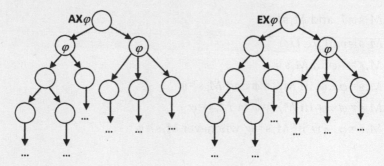

FIGURE 6.10
Illustration of semantics of **AX** and **EX**.

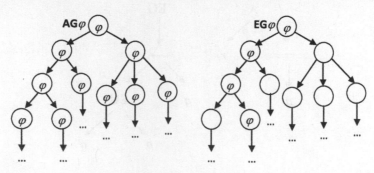

FIGURE 6.11
Illustration of semantics of **AG** and **EG**.

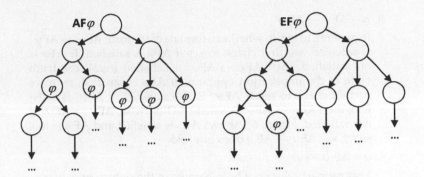

FIGURE 6.12
Illustration of semantics of **AF** and **EF**.

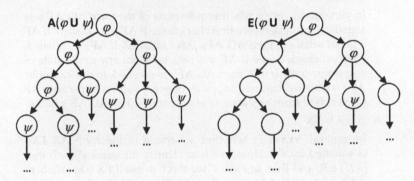

FIGURE 6.13
Illustration of semantics of **AU** and **EU**.

Example 6.4: CTL Formula Verification

In this example, we verify the system introduced in Example 6.1 the satisfaction relation, $M, s_1 \vDash \varphi$, against a few CTL specifications from the initial state s_1:

1. $\varphi = \mathbf{AX}\ p$

   $M, s_1 \vDash \mathbf{AX}\ p$ does not hold because s_3 does not have p.

2. $\varphi = \mathbf{AF}\ q$

   Fq is true in every path. Therefore, $M, s_1 \vDash \mathrm{AF}\ q$ holds.

3. $\varphi = \mathbf{EG}(\neg p \wedge r)$

   $(\neg p \wedge r)$ does not hold in every state of any single path. Therefore, $M, s_1 \vDash \mathbf{EG}(\neg p \wedge r)$ does not hold.

4. $\varphi = \mathbf{E}(p \wedge r\ \mathbf{U}\ q)$

   We check to see if there is a path such that $p \wedge r$ remains true until q is true. The rightmost path in Figure 6.2 satisfies this property. Therefore, $M, s_1 \vDash \mathbf{E}(p \wedge r\ \mathbf{U}\ q)$ holds.

5. $\varphi = \mathbf{AF}\ r \to \mathbf{AF}\ q$

   In general, to verify whether a formula of the form $\mathbf{AF}\psi \to \mathbf{AF}\psi'$ is satisfied, we first check to see if $\mathbf{AF}\psi$ is satisfied. If $\mathbf{AF}\psi$ is not satisfied, then $\mathbf{AF}\psi \to \mathbf{AF}\psi'$ is satisfied (recall the truth table of the implication operator). If $\mathbf{AF}\psi$ is satisfied, then we further check to see if $\mathbf{AF}\psi'$ is also satisfied. If $\mathbf{AF}\psi'$ is satisfied, then $\mathbf{AF}\psi \to \mathbf{AF}\psi'$ is satisfied. Otherwise, $\mathbf{AF}\psi \to \mathbf{AF}\psi'$ is not satisfied. Figure 6.2 shows $\mathbf{AF}\ r$ is satisfied and $\mathbf{AF}\ q$ is too, so $M, s_1 \vDash \mathbf{AF}\ r \to \mathbf{AF}\ q$ does not hold.

6. $\varphi = \mathbf{AF}\ (r \to q)$

   This property does not hold because in the rightmost path, $r \to q$ is violated in s_4. This also shows that $\mathbf{AF}\ (r \to q)$ and $\mathbf{AF}\ r \to \mathbf{AF}\ q$ are different.

7. $\varphi = \mathbf{AG}\ \mathbf{AF}\ r$

   In general, to verify whether a formula of the form $\mathbf{AG}\ \mathbf{AF}\ \psi$ is satisfied by a system, we first check to see if $\mathbf{AF}\ \psi$ is satisfied. If $\mathbf{AF}\ \psi$ is not satisfied, then $\mathbf{AG}\ \mathbf{AF}\psi$ is not satisfied. If $\mathbf{AF}\psi$ is satisfied, we then check to see if $\mathbf{AF}\ \psi$ is satisfied from any single state of any single path. If it is, then $\mathbf{AG}\ \mathbf{AF}\ \psi$ is satisfied. In this example, $\mathbf{AF}\ r$ is satisfied because r appears on every path. Also because $\mathbf{AF}\ r$ is satisfied from any state of every path, $M, s_1 \vDash \mathbf{AG}\ \mathbf{AF}\ q$ holds.

8. $\varphi = \mathbf{AX}\ \mathbf{EX}\ p$

   In general, to verify whether a formula of the form $\mathbf{AX}\ \mathbf{EX}\psi$ is satisfied by a system, we first identify all states s' such that $(s, s') \in R$, and then in each s' we check to see if $\mathbf{EX}\ \psi$ is satisfied. If $\mathbf{EX}\ \psi$ is satisfied in all s', then $\mathbf{AX}\ \mathbf{EX}\ \psi$ holds. In this example, s_2 and s_3 are the two next states of s_1. $\mathbf{EX}\ p$ is satisfied in s_2 but not in s_3. Therefore, $M, s_1 \vDash \mathbf{AX}\ \mathbf{EX}\ p$ does not hold.

9. $\varphi = \mathbf{EX}\ \mathbf{AX}\ r$

   In general, to verify whether a formula of the form $\mathbf{EX}\ \mathbf{AX}\psi$ is satisfied by a system, we first identify all states s' such that $(s, s') \in R$. Then from each s' we check to see if $\mathbf{AX}\psi$ is satisfied. As long as $\mathbf{AX}\psi$ is satisfied from one s', $\mathbf{EX}\ \mathbf{AX}\psi$ holds. In this example, s_2 and s_3 are the two next states of s_1. $\mathbf{AX}\ r$ is satisfied by both of them. Therefore, $M, s_1 \vDash \mathbf{EX}\ \mathbf{AX}\ r$ holds.

6.3.3 Equivalencies of CTL Formulas

Two CTL formulas φ and ψ are said to be *semantically equivalent*, denoted by $\varphi \equiv \psi$, if for all models M and all states s in M,

$$M, s \vDash \varphi \text{ iff } M, s \vDash \psi$$

In other words, two CTL formulas are equivalent if either both of them are satisfied or none of them are satisfied from any state of any Kripke structure. Listed below are a few equivalencies.

The two path quantifiers **A** and **E** are duals: $\neg\mathbf{A}\,\varphi \equiv \mathbf{E}\,\neg\varphi$, which results in the following equivalencies:

$$\neg\mathbf{AX}\,\varphi \equiv \mathbf{EX}\,\neg\varphi$$

$$\neg\mathbf{AF}\,\varphi \equiv \mathbf{EG}\,\neg\varphi$$

$$\neg\mathbf{EF}\,\varphi \equiv \mathbf{AG}\,\neg\varphi$$

Recall in LTL we have $\mathbf{F}\,\varphi \equiv \top\,\mathbf{U}\,\varphi$. In CTL we also have similar equivalencies:

$$\mathbf{AF}\,\varphi \equiv \mathbf{A}(\top\,\mathbf{U}\,\varphi)$$

$$\mathbf{EF}\,\varphi \equiv \mathbf{E}(\top\,\mathbf{U}\,\varphi)$$

Other equivalencies include:

$$\mathbf{AX}\,(\varphi \wedge \psi) \equiv \mathbf{AX}\,\varphi \wedge \mathbf{AX}\,\psi$$

$$\mathbf{EX}\,(\varphi \vee \psi) \equiv \mathbf{EX}\,\varphi \vee \mathbf{EX}\,\psi$$

$$\mathbf{AG}\,(\varphi \wedge \psi) \equiv \mathbf{AG}\,\varphi \wedge \mathbf{AG}\,\psi$$

$$\mathbf{EF}\,(\varphi \vee \psi) \equiv \mathbf{EF}\,\varphi \vee \mathbf{EF}\,\psi$$

$$\mathbf{AF\,AF}\,\varphi \equiv \mathbf{AF}\,\varphi$$

$$\mathbf{EF\,EF}\,\varphi \equiv \mathbf{EF}\,\varphi$$

$$\mathbf{AG\,AG}\,\varphi \equiv \mathbf{AG}\,\varphi$$

$$\mathbf{EG\,EG}\,\varphi \equiv \mathbf{EG}\,\varphi$$

6.3.4 System Property Specification

We use the elevator of five floors introduced in Section 6.2.5 as an example to show how we can use CTL formulas to formalize some requirements of a real system. We reuse the atomic propositions defined in Section 6.2.5.

1. The elevator should not move if the door is open.

$$\mathbf{AG}(\neg door\_closed \rightarrow \neg(direction\_up \vee direction\_down))$$

2. Whenever the door is open, it will eventually be closed.

$$\mathbf{AG}(\neg door\_closed \rightarrow \mathbf{AF}\,door\_closed)$$

3. When the elevator is traveling upward, it does not change its direction when there are passengers waiting to go to a higher floor. For example:

$$\mathbf{AG}\ ((floor\_1 \vee floor\_2 \vee floor\_3) \wedge direction\_up \wedge button\_4$$

$$\rightarrow \mathbf{A}(direction\_up\ \mathbf{U}\ floor\_4))$$

The above three formulas are adopted from their LTL forms by replacing **G**, **F**, and **U** with **AG**, **AF**, and **AU**, respectively. The property in formula 4 can only be coded in CTL.

4. It is always possible that the elevator moves up from the lowest floor without any stop on the way up until it reaches the highest floor.

$$\mathbf{AG}(floor\_1 \rightarrow \mathbf{E}\ (direction\_up \wedge door\_closed\ \mathbf{U}\ floor\_5))$$

6.3.5 LTL versus CTL

LTL and CTL are two of the most popular forms of temporal logic. LTL views time as a linear path extending to the future, while CTL views time as a branching-out structure. In general, LTL formulas are more intuitive and easier to understand. Using a combination of path quantifiers and temporal operators, CTL formulas are less intuitive and thus more error-prone in system property specification.

They overlap in their expressive powers. CTL allows explicit quantification over paths, which makes it more expressive than LTL in that regard. In fact, any CTL formula necessitating the operator **E** cannot be expressed in LTL.

On the other hand, there are also LTL formulas that cannot be expressed in CTL. One such a formula, for example, is formula $F\varphi \rightarrow F\psi$. It means that "all paths that have a state satisfying φ along them also have a state satisfying ψ along them." We might think that it is equivalent to the CTL formula $\mathbf{AF}\varphi \rightarrow \mathbf{AF}\psi$. Actually it is not, because this CTL formula means "if all paths have a state satisfying φ along them, then all paths have a state satisfying ψ along them." Let $\varphi = p \wedge r$ and $\psi = p \wedge q$, then the model shown in Figure 10.5 does not satisfy the LTL formula $F\varphi \rightarrow F\psi$, because φ is satisfied by the left child node of the initial state, but $F\psi$ is not satisfied along all path starting the left child node. However, the model does satisfy the CTL formula $\mathbf{AF}\varphi \rightarrow \mathbf{AF}\psi$ because $\mathbf{AF}\varphi$ is not satisfied.

Figure 6.14 shows the expressive powers of LTL and CTL.

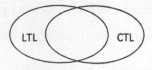

FIGURE 6.14
Expressive powers of LTL and CTL.

6.4 CTL*

A variation of CTL is CTL*. CTL* is a superset of CTL and LTL. Like CTL, CTL* is a branching-time logic. The difference is that each temporal operator in CTL is a combination of a path quantifier and linear-time operator, while CTL* freely combines path quantifiers and temporal operators.

There are two classes of CTL* formulas, namely *state formulas* and *path formulas*. The definitions of state formulas and path formulas are *mutually recursive*: the definition of one class depends on the definition of the other class. State formulas are evaluated in states and defined as follows:

$$\varphi ::= \top \mid \bot \mid p \mid (\neg\varphi) \mid (\varphi \wedge \varphi) \mid (\varphi \vee \varphi) \mid (\varphi \rightarrow \varphi)$$
$$\mid (\mathbf{A}\psi) \mid (\mathbf{E}\psi) \tag{6.3}$$

where p is any atomic proposition, and ψ is any path formula.

Path formulas are evaluated over paths and defined as follows:

$$\psi ::= \varphi \mid (\neg\psi) \mid (\psi \wedge \psi) \mid (\psi \vee \psi) \mid (\psi \rightarrow \psi)$$
$$\mid \mathbf{X}\psi) \mid \mathbf{G}\psi) \mid \mathbf{F}\psi \mid \psi\mathbf{U}\psi \tag{6.4}$$

where φ is any state formula.

Based on the above definitions, we can construct the following well-formed CTL* formulas:

1. $p \wedge \neg q$
2. $\mathbf{G}(p \wedge \neg q)$
3. $\mathbf{EG}(p \wedge \neg q)$
4. $\mathbf{AX}(p \vee r)$
5. $\mathbf{AFG}(p \wedge \neg q)$
6. $\mathbf{EX}(p \vee r) \ \mathbf{U} \ \mathbf{AFG}(p \wedge \neg q)$

Formula 1 is a propositional logic formula, and thus it is in both LTL and CTL. Formula 2 is an LTL formula, and its equivalent in CTL is $\mathbf{AG}(p \wedge \neg q)$. The third formula is a CTL formula but it is not in LTL. The forth formula is an CTL formula, and its equivalent in LTL is $\mathbf{X}(p \vee r)$. Formula 5 is in LTL, but it is not in CTL. The last formula is neither in LTL nor in CTL.

Both LTL and CTL are a restricted subset of CTL*. More specifically, LTL consists of formulas that have the form $\mathbf{A}\psi$, where ψ is a path formula in which the only state subformulas permitted are atomic propositions. $\mathbf{A}(\mathbf{FG}\ p)$ is such an example. CTL consists of formulas that permit only branching-time

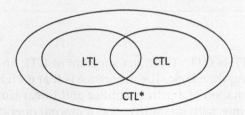

FIGURE 6.15
Expressive powers of LTL, CTL, and CTL*.

operators—each linear-time operator **G**, **F**, **X**, or **U** must be immediately preceded by a path quantifier. **AF(EG** p) is an example of CTL formulas. Figure 6.15 illustrates the relationship among the expressive powers of CTL*, LTL, and CTL.

The formal semantics of CTL* formulas are defined in terms of Kripke structures. If φ is a state formula, then the relation $M, s \models \varphi$ means that φ holds in the state s of the Kripke structure M. Assume that φ_1 and φ_2 are two state formulas and ψ is a path formula. Then the relation $M, s \models \varphi$ is defined as follows:

1. $s \models p$ iff $p \in L(s)$
2. $s \models \neg\varphi_1$ iff $s \not\models \varphi_1$
3. $s \models \varphi_1 \wedge \varphi_2$ iff $s \models \varphi_1$ and $s \models \varphi_2$
4. $s \models \varphi_1 \vee \varphi_2$ iff $s \models \varphi_1$ or $s \models \varphi_2$
5. $s \models \varphi_1 \rightarrow \varphi_2$ iff $s \models \varphi_2$ as long as $s \models \varphi_1$
6. $s \models A\psi$ iff for all paths π starting with s, $\pi \models \psi$
7. $s \models E\psi$ iff there is a path π starting with s such that $\pi \models \psi$

If ψ is a path formula, then the relation $M, \pi \models \psi$ means that ψ holds along the path π of the Kripke structure M. Assume that ψ_1 and ψ_2 are two path formulas and φ is a state formula. Then the relation $M, \pi \models \varphi$ is defined as follows:

1. $\pi \models \varphi$ iff $s \models \varphi$ and s is the first state of φ
2. $\pi \models \neg\psi_1$ iff s—ψ_1
3. $\pi \models \psi_1 \wedge \psi_2$ iff $\pi \models \psi_1$ and $\pi \models \psi_2$
4. $\pi \models \psi_1 \vee \psi_2$ iff $\pi \models \psi_1$ or $\pi \models \psi_2$
5. $\pi \models \psi_1 \rightarrow \psi_2$ iff $\pi \models \psi_2$ as long as $\pi \models \psi_1$
6. $\pi \models X\psi_1$ iff $\pi_2 \models \psi_1$
7. $\pi \models G\psi_1$ iff $\pi_i \models \psi_1$ for all $i \geq 1$
8. $\pi \models F\psi_1$ iff $\pi_i \models \psi_1$ for some $i \geq 1$
9. $\pi \models \psi_1 U\psi_2$ iff $\pi_i \models \psi_2$ for some $i \geq 1$ and $\pi_j \models \psi_2$ for all $j = 1, 2, ..., i-1$

Exercises

1. What is the difference between LTL and CLT in terms of modeling of time?

2. Can any LTL formula be specified by CTL? Can any CTL formula be specified by LTL?

3. What is the difference between CTL and CTL*?

4. Draw the parse trees of the following LTL formulas:
   a. $q \wedge Xp \rightarrow F(q \vee \neg r)$
   b. $(p \rightarrow \neg q) \wedge Xr \rightarrow GFq$
   c. $(p \, U \, Gq) \rightarrow (Fq \wedge Xr)$
   d. $X(p \rightarrow q) \, U \, (Fq \rightarrow Gr)$

5. Draw parse trees for the following CTL formulas:
   a. $EG(\neg p \wedge EXp) \rightarrow AFq$
   b. $A(\neg p \, U \, q) \rightarrow (AFq \wedge EG(p \vee r))$
   c. $E(\neg p \, U \, (AG \, q)) \vee AG \, (q \rightarrow r)$
   d. $AG \, A(\neg p \, U \, (q \wedge r))$

6. Consider the state transition system M illustrated in Figure 6.16. For each LTL formula φ listed below, decide if $M, s_1 \models \varphi$ holds.
   a. $p \vee q$
   b. $X \, (q \wedge r)$
   c. $(p \vee q) \wedge X \, q$
   d. $XX \, (p \vee q)$
   e. $XX \, (q \wedge r)$
   f. $F \, (p \vee \neg r)$
   g. $F \, (p \wedge q)$
   h. $G \, p$
   i. $G \, (p \vee r)$
   j. $p \, U \, r$
   k. $\neg q \, U \, (\neg p \wedge r)$
   l. $G \, (\neg p \rightarrow q)$
   m. $G \, (p \, U \, (q \wedge r))$
   n. $GF \, p$
   o. $G(p \rightarrow X \, r)$
   p. $F(q \rightarrow \neg r)$
   q. $GF \, (q \rightarrow r)$

    r. **F** $(p \wedge q) \rightarrow$ **F** $(q \wedge r)$

    s. **GF** $(p \wedge q) \rightarrow$ **GF** $\neg(q \vee r)$

    t. **G**$(p \rightarrow$ **X** $r)$

    u. **F**$(q \rightarrow \neg r)$

    v. **GF**$(p \rightarrow r)$

7. Consider the state transition system M illustrated in Figure 6.16. For each CTL formula φ listed below, decide if $M, s_1 \models \varphi$ holds.

    a. **AG** $(p \wedge r)$

    b. **AF** $(p \vee q)$

    c. **AX AX** p

    d. **EX AX** p

    e. **AF** $(q \wedge r)$

    f. **EF**$(q \wedge r)$

    g. **AG** $(p \wedge (\neg q \vee r))$

    h. **EG** $(p \wedge (q \vee \neg r))$

    i. **E**$(p$ **U** $(q \wedge r))$

    j. **A**$(p$ **U** $(q \wedge r))$

    k. **A**$(p$ **U EG** $q)$

    l. **E**$(p$ **U AG** $q)$

    m. **EG** $(p \rightarrow \neg q)$

    n. **EG AF** r

    o. **AG EF** r

    p. **EF** $(q \rightarrow \neg r)$

    q. **EG** $(p \rightarrow$ **X** $r)$

    r. **GF** $(r \rightarrow q)$

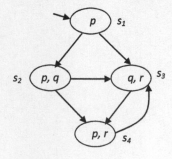

FIGURE 6.16
The state transition model for Problems 6 and 7.

8. Consider two LTL formulas $\varphi_1 = \mathbf{F}\,p$ and $\varphi_2 = \mathbf{FX}\,p$. Prove that these formulas are equivalent or exhibit a counterexample, that is, give a path such that one formula holds on the path and the other doesn't.

9. Consider two LTL formulas $\varphi_1 = p\,\mathbf{U}\,(q\,\mathbf{U}\,r)$ and $\varphi_2 = (p\,\mathbf{U}\,q)\,\mathbf{U}\,r$. Prove that these formulas are equivalent or exhibit a counterexample.

10. Construct counterexamples to prove the following pairs of formulas are not equivalent:

   a. $p \to q$ and $p\,\mathbf{U}\,q$

   b. $\mathbf{FG}\,p$ and $\mathbf{GF}\,p$

   c. $\mathbf{GX}\,p$ and $\mathbf{G}\,p$

11. Consider a traffic light. Write LTL formulas that formalize the following requirements

   a. The light must change colors in the following sequence: green, red, and yellow.

   b. The light must be in green color infinitely often.

   c. The light cannot change color from green to red right away.

12. Consider a microwave oven. It has three states: *door_open, ready,* and *heating*. Write CTL formulas that formalize the following requirements:

   a. The oven never heats while the door is open.

   b. When the oven is ready, one can start heating or open the door (to load or unload food).

   c. When the oven is heating, one can open the door, which automatically stops the heating.

7

Formal Verification by Model Checking

Model checking is an automatic verification technique for finite-state concurrent and reactive systems. It is developed to verify if assertions on a system are true or false. This is compared at program testing or simulation, which is to find out if there are bugs in a system. The aim of this chapter is to introduce the model checking technique. A model checking tool, NuSMV, and its associated system description language are also presented.

7.1 Introduction to Model Checking

Testing and simulation are the two most common approaches to ensure software correctness. Software testing involves the execution of a software component or system component to evaluate one or more properties of interest. This approach is very useful in practice, although it is clearly not possible to use it in highly critical systems if errors in the testing data could cause damage before actual deployment. Simulation is based on the process of modeling a real system with a set of mathematical formulas. It is, essentially, a program that allows the user to observe an operation through simulation without actually performing that operation. Simulation does not work directly on the real system, which is a big advantage over testing.

Both testing and simulation are widely applied in industrial applications. However, program testing or simulation can only show the presence of errors, never their absence. It is not possible, in general, to simulate or test all the possible scenarios or behaviors of a given system. In other words, those techniques are not exhaustive due to the high number of possible cases to be taken into account, and the failure cases may be among those not tested or simulated.

Model checking is an automatic verification technique for finite-state concurrent systems. It originated independently by the pioneering work of E. M. Clarke and E. A. Emerson and by J. P. Queille and J. Sifakis in the early 1980s. In model checking, digital circuits or software designs under study are modeled as state transition systems, and desired properties are specified

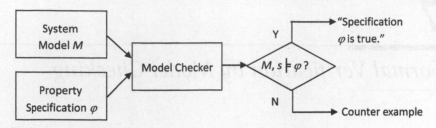

FIGURE 7.1
Model checking.

with temporal logic formulas. Verification is performed to find out whether the finite-state model meets the specifications. Verification is carried out by running model checking tools, which are also called *model checkers*.

Model checking technique is based on solid mathematics, logics, and computer science. It is a formal verification method and can be stated as follows: given a desired property, expressed as a temporal logic formula φ, and a structure M, decide if $M, s \models \varphi$, that is, if φ is satisfied by M from a given state s. If it is, the model checking tool or model checker will simply output something like a "yes"; If it is not, the tool will print out a counterexample of execution in which the property is violated. This is illustrated in Figure 7.1.

7.2 CTL Model Checking Algorithm

This section introduces the classic *labeling algorithm* for CTL formula model checking. Recall that in Chapter 6 we mentioned that **A** and **E** and **G** and **F** are duals, and $\mathbf{F}\,\varphi \equiv \top\,\mathbf{U}\,\varphi$. Moreover, $\top$ and $\bot$ are duals, $\wedge$ and $\vee$ are duals, and $\varphi \rightarrow \psi \equiv \neg\varphi \vee \psi$. Therefore, we can convert any CTL formula to an equivalent one that has only the logic constant $\bot$, Boolean operators $\neg$ and $\wedge$, and temporal operators **AF**, **EU**, and **EX**. We say these operators together form an *adequate set* of CTL operators.

7.2.1 The Labeling Algorithm

Instead of answering the question if $M, s \models \varphi$ holds, the labeling algorithm computes the set of all states s' that has $M, s' \models \varphi$, and then checks to see if s is in the set. Before applying the algorithm, we convert φ to an equivalent that has only operators from the adequate set. Next, we label the states of M with the subformulas of φ that are satisfied in these states, starting with the innermost and smallest subformulas and working outward toward φ.

Suppose that ψ is a subformula of φ and all states that satisfy its immediate subformulas have been labeled. The labeling of states in terms of ψ depends on what ψ is and its structure. Consider the following cases:

- $\psi = \bot$: No states are labeled.
- $\psi = p$: An atomic proposition: label all states s if $p \in L(s)$.
- $\psi = \psi_1 \wedge \psi_2$: Label all states that are already labeled with ψ_1 and ψ_2.
- $\psi = \neg\psi_1$: Label all states that are not labeled with ψ_1.
- $\psi = \mathbf{EX}\,\psi_1$: Label any state with $\mathbf{EX}\,\psi_1$ if the state has at least one successor labeled with ψ_1. This is illustrated in Figure 7.2.
- $\psi = \mathbf{AF}\psi_1$: For each state that is labeled with ψ_1, we label it with $\mathbf{AF}\psi_1$. After that, we identify states that each has all next states labeled with $\mathbf{AF}\psi_1$, and label all such states with $\mathbf{AF}\psi_1$. We repeat this step until there is no state to label. This labeling process is illustrated in Figure 7.3.
- $\psi = \mathbf{E}(\psi_1\ \mathbf{U}\ \psi_2)$: For each state that is labeled with ψ_2, we label it with $\mathbf{E}(\psi_1\ \mathbf{U}\ \psi_2)$. After that, for each state that is labeled with ψ_1 and also has at least one successor labeled with $\mathbf{E}(\psi_1\ \mathbf{U}\ \psi_2)$, we label it with $\mathbf{E}(\psi_1\ \mathbf{U}\ \psi_2)$. We repeat this step until there is no state to label. This labeling process is illustrated in Figure 7.4.

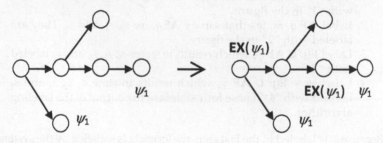

FIGURE 7.2
Labeling of $\mathbf{EX}(\psi_1)$.

FIGURE 7.3
Labeling of $\mathbf{AF}\psi_1$.

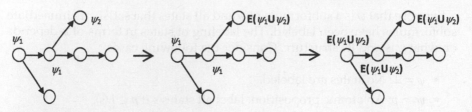

FIGURE 7.4
Labeling of $E(\psi_1 \ U \ \psi_2)$.

Example 7.1: Labeling Algorithm

Given the Kripke structure in Figure 7.5, verify the property

$$\varphi = r \rightarrow E(p \ \mathbf{U} \ \mathbf{AF} \ q)$$

Solution

First, we convert it to an equivalent without the implication operator:

$$\varphi' = \neg r \lor E(p \ \mathbf{U} \ \mathbf{AF} \ q)$$

Its parse tree is shown in Figure 7.6. The labeling process takes the following steps:

1. Label $\neg r$. States that satisfy $\neg r$ are s_1, s_4, and s_5. They are labeled with "1" in the figure.
2. Label $\mathbf{AF} \ q$. States that satisfy $\mathbf{AF} \ q$ are s_2, s_4, and s_5. They are labeled with "2" in the figure.
3. Label $E(p \ \mathbf{U} \ \mathbf{AF} \ q)$, which results in states s_1, s_2, s_4, and s_5 labeled with "3."
4. Label $\neg r \lor E(p \ \mathbf{U} \ \mathbf{AF} \ q)$, which results in state s_1, s_2, s_4 and s_5 labeled with "4." These four states are the output of the labeling algorithm.

Because s_1 is labeled in the last step, the formula is satisfied by the system.

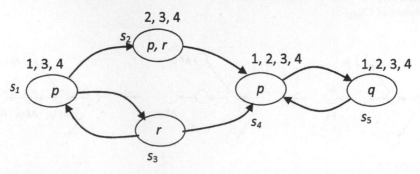

FIGURE 7.5
Kripke structure of Example 7.1.

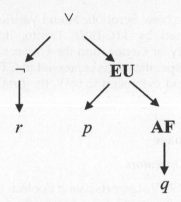

FIGURE 7.6
Parse tree of the formula $\neg r \vee E(p\ U\ AF\ q)$.

7.2.2 State Explosion Issues in Model Checking

The labeling algorithm is simple to understand and can be fully automated. However, it also experiences some shortcomings. Because the algorithm involves exhaustive search of the model state space to determine whether or not a formula is satisfied, it suffers from the well-known *state explosion* problem. As we know, the state space of a system grows experientially with the number of concurrent processes. For example, if there are m processes in a system and each process has n states, then the total number of states is m^n. On the other hand, when we add a new logic variable to a model, its state space is doubled in the worst case.

There are several approaches that tackle the state explosion problem. One approach is *symbolic model checking* with *binary decision diagrams* (BDDs). In this approach, a set of states is represented by a BDD instead of by listing each state individually. The BDD representation is often exponentially smaller in practice. Model checking with BDDs is performed using a fixed point algorithm.

7.3 The NuSMV Model Checking Tool

There is a long list of model checking tools developed to support system property verification. SPIN, NuSMV, FDR2, CADP, and ProB are some examples. They differ in terms of property specification languages and modeling languages. In this book, we only introduce the NuSMV model checker.

NuSMV is short for the New Symbolic Model Verifier. It is an open-source product jointly developed by ITC-IRST, Trento, Italy, Carnegie Mellon University, the University of Genoa, and the University of Trento. NuSMV supports the analysis of specifications expressed in CTL and LTL. NuSMV is a reimplementation of and extension to SMV, the first model checker based on BDD.

7.3.1 Description Language

7.3.1.1 Data Types and Operators

The input language of NuSMV supports data of Boolean type, enumerative type, bounded integers, and finite arrays. For example, the following code segment defines a Boolean variable cond, an enumerative variable status that takes values from {ready, busy, waiting, stopped}, an integer variable num that is bounded from 1 to 10, and an array arr of Booleans indexed from 0 to 10.

```
VAR
  cond : Boolean;
  status : {ready, busy, waiting, stopped};
  num : 1..10;
  arr : array 0..10 of Boolean;
```

There are several types of operators in the language. Logic operators include ! (not), & (and), | (or), xor (exclusive or), -> (implication), and <-> (if and only if). Relations operators include > (greater than), < (less than), ≥ (greater than or equal to), ≤ (less than or equal to), = (equality), and != (inequality). Arithmetic operators include + (addition), − (unary negation or subtraction), * (multiplication), and / (division). Set operators include in (set inclusion) and union (set union). The semantics of these operators are consistent with their uses in other computer science subjects.

7.3.1.2 Single-Module SMV Program

An SMV program is typically broken down into *modules* that can be composed and reused. Each module defines a finite state machine. It declares variables, defines initial values of variables, and specifies how the values change in each step.

Figure 7.7 shows a program of single module. The section starting with the keyword VAR declares two variables, in which request is Boolean while state is enumerative. The state variables decide the state space of the model.

The rest of the code is the assignment section, starting with the keyword ASSIGN. The first part of the assignment section assigns an initial value to each variable. The keyword init is used to describe the initial value of a variable. The syntax is

```
MODULE main
  VAR
      request : boolean;
      state : {ready, busy};
  ASSIGN
      init(state) := ready;
      next(state) :=
          case
              state = ready & request    : busy;
              TRUE                        : {ready, busy};
          esac;
```

FIGURE 7.7
An single-module SMV program.

```
init(<variable>) := <expression>;
```

where <expression> must evaluate to values in the domain of <variable>. If the initial value of a variable is not specified, then the variable can take any value in its domain as its initial value. In this example, the variable state is initialized to ready, but the variable request is not initialized, meaning its initial value can be either TRUE or FALSE. Therefore, there are two initial states in this model:

```
(request = TRUE, state = ready)
(request = FALSE, state = ready)
```

The second part of the assignment section specifies state transitions. It assigns values with the keyword next, which describes how the value of the variable changes in one step. The syntax of the next statement is

```
next(<variable>) := <next_expression>;
```

where <next_expression> must evaluate to values in the domain of <variable>. <next_expression> depends on "current" and "next" variables. For example,

```
next(x) := x | y;
next(y) := y & next(x);
```

Here the value of x in the next state, that is, the next value of x is the disjunction of its current value and the current value of y. The next y is the conjunction of current y and next X. If the next value of a variable is not specified, then the variable may take any value in its domain in the next state. In this single-module program, the change of state is specified with a case statement, but not of request.

A case statement assigns the value of the variable associated with each case condition when it is true; TRUE is for all default cases. In general, it is written as

```
case
    cond1: expr1;
    cond2: expr2;
    ...
    TRUE: exprN;
esac;
```

Let us consider all possible transitions from the first initial state. Based on the first condition of the case statement, the next value of state should be busy. The next value of request is not specified and thus it can be either TRUE or FALSE. Therefore, the system can transit from the initial state (request = TRUE, state = ready) to either (request = TRUE, state = busy) or (request = FALSE, state = busy).

Now consider the second initial state. The first condition of the case statement for state is not satisfied, so its next value can be either ready or busy, as defined by the default case. The next value of request again can be either TRUE or FALSE. Therefore, the system can transit from the initial state (request = TRUE, state = ready) to a state of any combination of values of request and state. The state transition model of the program is fully depicted in Figure 7.8.

7.3.1.3 Multi-Module SMV Program

An SMV program can consist of more than one module. In each SMV specification there must be a module main. It is the topmost module. All other modules are instantiated in main or other parent modules. The instantiation is performed inside the VAR declaration of the parent module. All the variables declared in a module instance are visible in the module in which it has been instantiated via the dot notation.

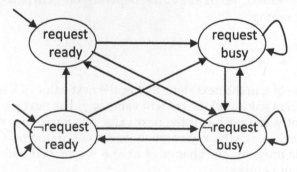

FIGURE 7.8
The transition system corresponding to the SMV program in Figure 7.7.

```
MODULE counter_cell(carry_in)
   VAR
      value : boolean;

   ASSIGN
      init(value) := FALSE;
      next(value) := value xor carry_in;
   DEFINE carry_out := value & carry_in;

MODULE main
   VAR
      bit0 : counter_cell(TRUE);
      bit1 : counter_cell(bit0.carry_out);
      bit2 : counter_cell(bit1.carry_out);
```

FIGURE 7.9
SMV program of a three-bit counter.

The program listed in Figure 7.9 has two modules: `main` and `counter_cell`. It is a model of a three-bit binary counter circuit. As the name indicates, the entry module of the program is `main`. The `main` module simply initiates three instances of the `counter_cell` module, named `bit0`, `bit1`, and `bit2`. The `counter_cell` module has a parameter `carry_in`. For example, the `carry_in` of `bit1` is `bit0.carry_out`. Note that an expression of the form `a.b` denotes the component `b` of module `a`, just as if the module `a` were a data structure in a standard programming language. Hence, the D of module `bit1` is the `carry_out` of module `bit0`. `DEFINE` is used to define C-like "macros"; defined variables are not real variables in that they do not increase the state space. The operator `xor` means "exclusive or," a Boolean operator working on two variables that has the value of one (TRUE) if one but not both of the variables has a value of one (TRUE).

The initial state of the model is

```
(bit0.value = FALSE, bit1.value = FALSE, bit2.value = FALSE)
```

or briefly described as 000. It transitions to 001, 010, … , all the way to 111 and then repeats. The details of the first three states are listed in Table 7.1. The result of each state is recorded in `value`.

7.3.1.4 Asynchronous Systems

The previous two programs describe synchronous systems, wherein each module or module instance the assignment statements are taken into account in parallel and simultaneously at each "clock tick." NuSMV allows for asynchronous system modeling. It is possible to define a collection of parallel processes, whose actions are interleaved, following an asynchronous model of concurrency.

TABLE 7.1

First Two State Transitions of the Program Listed in Figure 7.9

	State 0			State 1			State 2		
	bit0	bit1	bit2	bit0	bit1	bit2	bit0	bit1	bit2
Carry_in	T	F	F	T	T	F	T	T	F
Value	F	F	F	T	F	F	F	T	F
Next(value)	T	F	F	F	T	F	T	F	F
Carry_out	F	F	F	T	F	F	F	T	F

Figure 7.10 shows an SMV program that represents a ring of three asynchronous inverting gates. Here, the keyword process specifies asynchronous module instances. Each time the global clock ticks, only one of the three inverter instances is randomly chosen to execute, and the values of variables of other instances remain unchanged. Since the system is not forced to eventually choose a given process to execute, it is possible that the output of a given gate may remain constant forever, regardless of its input. Thus, a statement

```
FAIRNESS
  running
```

is added to the end of the inverter module to force every instance of the inverter to execute infinitely often.

7.3.2 Specifications

Specifications can be added in any module of an SMV program. Each property is verified separately. NuSMV supports specifications in LTL and CTL. A property in LTL is specified with the keyword LTLSPEC:

```
MODULE inverter(input)
  VAR
    output : boolean;
  ASSIGN
    init(output) := FALSE;
    next(output) := !input;

MODULE main
  VAR
    gate1 : process inverter(gate3.output);
    gate2 : process inverter(gate1.output);
    gate3 : process inverter(gate2.output);
  FAIRNESS
    running
```

FIGURE 7.10
SMV program of an inverter ring.

```
LTLSPEC <ltl_expr>
```

where `<ltl_expr>` is an LTL formula coded in NuSMV:

```
ltl_expr::
   simple_expr;; a simple boolean expression
   | "("ltl_expr ")"
   | "!" ltl_expr;; logical not
   | ltl_expr "&" ltl_expr;; logical and
   | ltl_expr "|" ltl_expr;; logical or
   | ltl_expr "xor" ltl_expr;; logical exclusive or
   | ltl_expr "->"ltl_expr;; logical implies
   | ltl_expr "<->"ltl_expr;; logical equivalence;;
   | "X" ltl_expr;; next state
   | "G" ltl_expr;; globally
   | "F" ltl_expr;; finally
   | ltl_expr "U" ltl_expr;; until
   | ltl_expr "V" ltl_expr;; releases
```

For example, we can add

```
LTLSPEC F (bit0.value & bit1.value & bit2.value)
```

to the end of the program listed in Figure 7.9. This specification checks whether the property that eventually the counters outputs 111 holds. For the same program, we can also add

```
LTLSPEC G F bit2.value
```

which checks whether the third bit becomes true infinitely often.

A property in CTL is specified with the keyword SPEC:

```
SPEC <ctl_expr>
```

where `<ctl_expr>` is a CTL formula coded in NuSMV:

```
ctl_expr ::
   simple_expr ;; a simple boolean expression
   | "(" ctl_expr ")"
   | "!" ctl_expr ;; logical not
   | ctl_expr "&" ctl_expr ;; logical and
   | ctl_expr "|" ctl_expr ;; logical or
   | ctl_expr "xor" ctl_expr ;; logical exclusive or
   | ctl_expr "->" ctl_expr ;; logical implies
   | ctl_expr "<->" ctl_expr ;; logical equivalence
   | "EG" ctl_expr ;; exists globally
   | "EX" ctl_expr ;; exists next state
   | "EF" ctl_expr ;; exists finally
   | "AG" ctl_expr ;; forall globally
   | "AX" ctl_expr ;; forall next state
   | "AF" ctl_expr ;; forall finally
   | "E" "[" ctl_expr "U" ctl_expr "]" ;; exists until
   | "A" "[" ctl_expr "U" ctl_expr "]" ;; forall until
```

For example, we can add

```
SPEC EX gate3.output
SPEC EX gate1.output -> EX gate2.output
SPEC EF ((!gate1.output) & (!gate2.output) & gate3.output)
```

to the end of the program listed in Figure 7.10.

In addition to properties specified in LTL or CTL formulas, NuSMV can also verify *invariant* properties. An invariant is a propositional property that must always hold and is specified using the keyword INVARSPEC:

```
INVARSPEC <expression>
```

For example, for the program listed in Figure 7.7, we add the following specification:

```
INVARSPEC state in {ready, busy}
```

to check to see if the variable state always has a legal value.

7.3.3 Running NuSMV

NuSMV can be used either interactively or in batch mode. To check a model against a set of specifications, we write the specification and system description in a file with the .smv extension, and type the command

```
NuSMV <file_name>.smv
```

NuSMV will check each specification automatically, informing whether it is satisfied or producing a trace (when possible) to demonstrate its violation. For example, we save the counter program with the two LTL specifications we mentioned earlier to a file named counter.smv and run the command

```
NuSMV counter.smv
```

We will get the following result:

```
*** This version of NuSMV is linked to the MiniSat SAT solver.
*** See http://minisat.se/MiniSat.html
*** Copyright (c) 2003-2006, Niklas Een, Niklas Sorensson
*** Copyright (c) 2007-2010, Niklas Sorensson
-- specification F ((bit0.carry_out & bit1.carry_out) & bit2.
   carry_out) is true
-- specification G (F bit2.value) is true
```

If we modify the second specification to

```
LTLSPEC G X bit2.value
```

and run the command again, we get

```
*** This version of NuSMV is linked to the MiniSat SAT solver.
*** See http://minisat.se/MiniSat.html
*** Copyright (c) 2003-2006, Niklas Een, Niklas Sorensson
*** Copyright (c) 2007-2010, Niklas Sorensson
-- specification F ((bit0.value&bit1.value)&bit2.value) is true
-- specification G (X bit2.carry_out) is false
-- as demonstrated by the following execution sequence
Trace Description: LTL Counterexample
Trace Type: Counterexample
  -- Loop starts here
  -> State: 1.1 <-
    bit0.value = FALSE
    bit1.value = FALSE
    bit2.value = FALSE
    bit0.carry_out = FALSE
    bit1.carry_out = FALSE
    bit2.carry_out = FALSE
  -> State: 1.2 <-
    bit0.value = TRUE
    bit0.carry_out = TRUE
  -> State: 1.3 <-
    bit0.value = FALSE
    bit1.value = TRUE
    bit0.carry_out = FALSE
  -> State: 1.4 <-
    bit0.value = TRUE
    bit0.carry_out = TRUE
    bit1.carry_out = TRUE
  -> State: 1.5 <-
    bit0.value = FALSE
    bit1.value = FALSE
    bit2.value = TRUE
    bit0.carry_out = FALSE
    bit1.carry_out = FALSE
  -> State: 1.6 <-
    bit0.value = TRUE
    bit0.carry_out = TRUE
  -> State: 1.7 <-
    bit0.value = FALSE
    bit1.value = TRUE
    bit0.carry_out = FALSE
```

```
 -> State: 1.8 <-
    bit0.value = TRUE
    bit0.carry_out = TRUE
    bit1.carry_out = TRUE
    bit2.carry_out = TRUE
 -> State: 1.9 <-
    bit0.value = FALSE
    bit1.value = FALSE
    bit2.value = FALSE
    bit0.carry_out = FALSE
    bit1.carry_out = FALSE
    bit2.carry_out = FALSE
NuSMV >
```

The result says the second specification is false and it prints out a counterexample of execution that violates the specification. The counterexample lists all states step-by-step until they repeat.

NuSMV supports simulation that allows users to explore the possible executions (referred to *traces* from now on) of an SMV model. A simulation session is started interactively from the system prompt as follows

```
system_prompt> NuSMV -int <file>.smv
NuSMV> go
NuSMV>
```

The next step is to pick a state from the initial states to start a new trace. To pick a state randomly, type command

```
NuSMV> pick_state -r
```

Subsequent states in the simulation can be picked using the simulate command. For example, we can type the command

```
NuSMV> simulate -r -k 5
```

to randomly simulate 5 steps of a trace. To show the trace with states, use commands

```
NuSMV> show_trace -t
NuSMV> show_trace -v
```

Below is a screenshot of NuSMV simulation:

```
C:\...\NuSMV-2.6.0-win64\bin>nusmv -int counter.smv
...
NuSMV > go
NuSMV > pick_state -r
NuSMV > simulate -r -k 3
******** Simulation Starting From State 2.1 ********
```

```
NuSMV > show_traces -t
There are 2 traces currently available.
NuSMV > show_traces -v
    <!-- ################ Trace number: 2 ################ -->
Trace Description: Simulation Trace
Trace Type: Simulation
  -> State: 2.1 <-
    bit0.value = FALSE
    bit1.value = FALSE
    bit2.value = FALSE
    bit0.carry_out = FALSE
    bit1.carry_out = FALSE
    bit2.carry_out = FALSE
  -> State: 2.2 <-
    bit0.value = TRUE
    bit1.value = FALSE
    bit2.value = FALSE
    bit0.carry_out = TRUE
    bit1.carry_out = FALSE
    bit2.carry_out = FALSE
  -> State: 2.3 <-
    bit0.value = FALSE
    bit1.value = TRUE
    bit2.value = FALSE
    bit0.carry_out = FALSE
    bit1.carry_out = FALSE
    bit2.carry_out = FALSE
  -> State: 2.4 <-
    bit0.value = TRUE
    bit1.value = TRUE
    bit2.value = FALSE
    bit0.carry_out = TRUE
    bit1.carry_out = TRUE
    bit2.carry_out = FALSE
    bit2.carry_out = FALSE
NuSMV >
```

For details of how to use the NuSMV tool, please read the latest version of the NuSMV tutorial that can be downloaded from the official NuSMV website.

7.4 Example: The Ferryman Puzzle

In this example, we try to solve the classic ferryman puzzle with the NuSMV model checker. The puzzle is as follows: A ferryman has to bring a goat, a wolf, and a cabbage from one bank of a river to the opposite bank. The ferryman can cross the river either alone or with exactly one of these three passengers. At any time, either the ferryman should be on the same bank as the goat, or the goat should be alone on a bank. Otherwise, the goat will eat the cabbage or the wolf will eat the goat.

To model the system with the NuSMV program, we ignore the boat, as it is always with the ferryman. The four agents, the ferryman, wolf, goat, and cabbage, are modeled with four Boolean variables, as they each can have two values: *false* (on the initial bank of the river) and *true* (on the destination bank of the river). We model all possible behaviors in the program and ask if a trace indicating that the ferryman takes all of the three passengers to the other bank safely exists. The program is listed in Figure 7.11. In the program, we use the variable `carry` to indicate which passenger the ferryman will take with him to cross the river. If the ferryman travels alone, then `carry` takes the value of 0.

In the `ASSIGN` section of the program, the statement

```
next(ferryman):= !ferryman;
```

means that the ferryman must cross the river in each step. The `next(carry)` statement shows that the ferryman can take any passenger that is on the same bank of the river as he is, or take nothing with him (`union 0` at the end of the statement). The `next(goat)` statement says if the goat and ferryman are on the same bank and the goat is chosen to cross the river with the ferryman, then the goat will be on the other bank in the next state. Otherwise, the goat will stay on the same bank. The `next(cabbage)` and `next(wolf)` statements are similar to the `next(goat)` statement.

The LTL specification states that for all execution paths, if the goat and cabbage are on the same bank, or the goat and wolf are on the same bank, then the goat must be with the ferryman, and this is true until the ferryman and all of his passengers are on the destination bank. The program produces the following output:

```
C:\...\NuSMV-2.6.0-win64\bin>nusmv ferryman.smv
...
-- specification ((((goat = cabbage | goat = wolf) -> goat =
   ferryman) U (((cabbage & goat) & wolf) & ferryman)) is false
-- as demonstrated by the following execution sequence
Trace Description: LTL Counterexample
Trace Type: Counterexample
  -- Loop starts here
  -> State: 1.1 <-
     ferryman = FALSE
     goat = FALSE
     cabbage = FALSE
     wolf = FALSE
     carry = 0
  -> State: 1.2 <-
     ferryman = TRUE
  -> State: 1.3 <-
     ferryman = FALSE
```

```
MODULE main
  VAR
    ferryman : boolean;
    goat     : boolean;
    cabbage  : boolean;
    wolf     : boolean;
    carry    : {g, c, w, 0};

  ASSIGN
    init(ferryman) := FALSE;
    init(goat)     := FALSE;
    init(cabbage)  := FALSE;
    init(wolf)     := FALSE;
    init(carry)    := 0;

    next(ferryman) := !ferryman;

    next(carry) :=
      case
        (ferryman = goat) : g;
        TRUE              : 0;
      esac union
      case
        (ferryman = cabbage): c;
        TRUE                : 0;
      esac union
      case
        (ferryman = wolf)   : w;
        TRUE                : 0;
      esac union 0;

    next(goat) :=
      case
        (ferryman = goat) & (next(carry) = g)
                                 : next(ferryman);
        TRUE                     : goat;
      esac;

    next(cabbage) :=
      case
        (ferryman = cabbage) & (next(carry) = c)
                                 : next(ferryman);
        TRUE                     : cabbage;
      esac;

    next(wolf) :=
      case
        (ferryman = wolf) & (next(carry) = w)
                                 : next(ferryman);
        TRUE                     : wolf;
      esac;

  LTLSPEC
  ((goat=cabbage |goat = wolf) -> goat = ferryman)

        U (cabbage & goat & wolf & ferryman)
```

FIGURE 7.11
SMV program of the ferryman puzzle.

The result shows that the property is false, which is correct, because there are plenty of "execution paths" that violate this property. The program prints out one example, in which the ferryman crosses the river alone first (state 1.2), and then he travels back (state 1.3), and he keeps crossing the river this way.

Because what we are looking for is an execution path such that the LTL specification is true, we can verify an opposite specification, which is

```
LTLSPEC
!(((goat=cabbage |goat = wolf) -> goat = ferryman)
        U (cabbage & goat & wolf & ferryman))
```

If the puzzle has a solution, then the above specification won't be true on all execution paths. In that case, the trace of a counterexample will be printed on running the program, which is the solution that we want. The result from running the program with the new specification is listed below, in which states 1.1 through 1.8 form the trace of the solution, while states 1.9 through 1.15 form the trace that the ferryman takes all passengers back from the destination bank to the original bank safely.

```
C:\...\NuSMV-2.6.0-win64\bin>nusmv ferryman.smv
...
-- specification !(((goat = cabbage | goat = wolf) -> goat =
   ferryman) U (((cabbage & goat) & wolf) & ferryman)) is false
-- as demonstrated by the following execution sequence
Trace Description: LTL Counterexample
Trace Type: Counterexample
  -- Loop starts here
  -> State: 1.1 <-
     ferryman = FALSE
     goat = FALSE
     cabbage = FALSE
     wolf = FALSE
     carry = 0
  -> State: 1.2 <-
     ferryman = TRUE
     goat = TRUE
     carry = g
  -> State: 1.3 <-
     ferryman = FALSE
     carry = 0
  -> State: 1.4 <-
     ferryman = TRUE
     wolf = TRUE
     carry = w
  -> State: 1.5 <-
     ferryman = FALSE
     goat = FALSE
     carry = g
```

```
-> State: 1.6 <-
   ferryman = TRUE
   cabbage = TRUE
   carry = c
-> State: 1.7 <-
   ferryman = FALSE
   carry = 0
-> State: 1.8 <-
   ferryman = TRUE
   goat = TRUE
   carry = g
-> State: 1.9 <-
   ferryman = FALSE
   wolf = FALSE
   carry = w
-> State: 1.10 <-
   ferryman = TRUE
   carry = 0
-> State: 1.11 <-
   ferryman = FALSE
   cabbage = FALSE
   carry = c
-> State: 1.12 <-
   ferryman = TRUE
   carry = 0
-> State: 1.13 <-
   ferryman = FALSE
   goat = FALSE
   carry = g
-> State: 1.14 <-
   ferryman = TRUE
   carry = 0
-> State: 1.15 <-
   ferryman = FALSE
```

Exercises

1. Consider the Kripke structure illustrated in Figure 7.12. Verify the following properties using the labeling algorithm presented in Section 7.2.

   a. $p \wedge \mathbf{EX}\, q$

   b. $\mathbf{AF}\,(p \vee q)$

   c. $\neg r \wedge \mathbf{EF}\,(\neg p \wedge q)$

   d. $r \rightarrow \mathbf{AX}\, p$

   e. $\mathbf{E}(r\, \mathbf{U}\, \mathbf{AF}\, q)$

   f. $p \rightarrow \mathbf{E}(r\, \mathbf{U}\, \mathbf{AF}\, q)$

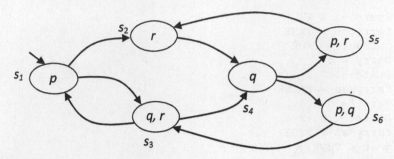

FIGURE 7.12
The Kripke structure for Problem 1.

2. Consider the state transition system *M* illustrated in Figure 7.13.

   a. Write the SMV program of the model.

   b. For each LTL or CTL formula φ listed below, verify if $M, s_1 \models \varphi$ holds by running the program.

      i. $(p \vee q) \wedge \mathbf{X} r$

      ii. $\mathbf{F}(p \wedge q)$

      iii. $\mathbf{FG} q$

      iv. $\mathbf{GF} q$

      v. $\mathbf{X} q \rightarrow (p \vee q) \mathbf{U} r$

      vi. $\mathbf{AX} r$

      vii. $\mathbf{EX} p$

      viii. $\mathbf{EG} q$

      ix. $\mathbf{EF\,AG} r$

      x. $\mathbf{A}(p \mathbf{U} r)$

      xi. $\mathbf{EX} r \rightarrow \mathbf{E}((p \vee q) \mathbf{U} q)$

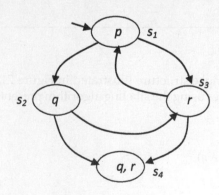

FIGURE 7.13
The state transition model for Problem 2.

3. Apply the labeling algorithm to verify all properties in Problem 2.

4. Table 7.1 listed the first two steps of state transitions of the three-bit counter system. Please list the details of the next six steps.

5. The SMV code listed in Figure 7.14 models the mutual exclusion feature of two processes. Each process can be in one of the three states: *idle, ready* (ready to enter the critical section), and *critical* (in the critical section). The properties under concern are as follows:

   Safety: at any moment, at most one task can be in its critical section.

   Liveness: whenever a task requests to enter its critical section, the request will eventually be granted.

   Fairness: if a task makes infinitely often requests to enter its critical section, it will enter its critical section infinitely often.

   a. Run the program and verify the properties of safety and liveness.

   b. Code the fairness property in LTL and verify it.

   c. Draw the state transition diagram of the program.

   d. Modify the program in Figure 7.14 to include a third process pr2 such that all the three properties are satisfied.

6. Figure 7.15 shows the control of a microwave oven in terms of its working status.

   a. Write an SMV program that implements the state transition system.

   b. Some fundamental requirements over the oven control include the following:

      i. The oven cannot heat unless the door is closed.

      ii. No matter what status the oven is in, it will be heating eventually.

   Code the two properties in CTL and verify them with the SMV program.

7. Consider a digital combination lock as illustrated in Figure 7.16. On its keypad are five digital input buttons for numbers 1–5, a reset button R, and a display that shows the number of keys pressed since the last reset. The lock code is a sequence of 4 digits. At the initial state, a 0 is displayed. When a number key is pressed, the display is incremented by 1, regardless of the input. Upon the fourth key press, if the input sequence is correct, the display will change to a 0, unlocking the system. If an incorrect sequence is entered, the display will change to an E upon the fourth key press, and the user must then press the R button to reset the lock to its initial state. The user can also press the R button at any time to reset the lock. Assume the code is set to 2019.

   a. Implement in NuSMV a system that encodes the behavior of the combination lock.

```
MODULE main
  VAR
    turn: {0, 1};
    pr0: process prc(pr1.control, turn, 0);
    pr1: process prc(pr0.control, turn, 1);

  ASSIGN
    init(turn) := 0;

    -- safety
  SPEC AG !((pr0.control = critical)&(pr1.control = critical))
    -- liveness
  SPEC AG ((pr0.control = ready) -> AF(pr0.control = critical))
  SPEC AG ((pr1.control = ready) -> AF(pr1.control = critical))


MODULE prc(other_control, turn, ID)
  VAR
    control: {idle, ready, critical};

  ASSIGN
    init(control) := idle;

    next(control) :=
      case
        (control = idle) : {ready, idle};
        (control = ready)&(other_control = idle): critical;
        (control = ready)&(other_control = ready)
                  &(turn = ID)    : critical;
        (control = critical)    : {critical, idle};
        TRUE                     : control;
      esac;

    next(turn) :=
      case
        (turn = ID)&(control = critical) : (turn + 1) mod 2;
        TRUE : turn;
      esac;

  FAIRNESS running;
  FAIRNESS !(control = critical);
```

FIGURE 7.14
SMV program of mutual exclusion.

b. Verify the correctness of the system by specifying and checking the following properties:

   i. The combination lock can always be unlocked.

   ii. The combination lock can be in the initial state infinitely often.

   iii. When the combination lock is in the E state, it will remain in the E state until the R button is pressed.

   iv. It is impossible that the combination lock can be unlocked from the E state within four steps.

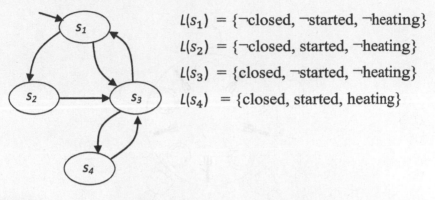

$L(s_1) = \{\neg closed, \neg started, \neg heating\}$

$L(s_2) = \{\neg closed, started, \neg heating\}$

$L(s_3) = \{closed, \neg started, \neg heating\}$

$L(s_4) = \{closed, started, heating\}$

FIGURE 7.15
State transition system of a microwave oven.

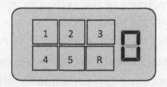

FIGURE 7.16
Keypad of a digital combination lock.

8. Consider the classic philosopher dining problem. As illustrated in Figure 7.17, five silent philosophers sit at a round table with bowls of spaghetti. There are also forks on the table, each between two adjacent philosophers. However, the spaghetti is of a particularly slippery type and a philosopher can only eat spaghetti when he has both left and right forks. The philosophers have agreed on the following protocols to grab the forks: Initially, they think about philosophy. When one gets hungry, he takes the fork on his left-hand side first, and then takes the one on his right-hand side, and then starts eating. He returns the two forks simultaneously to the table when he finishes eating and gets back to thinking about philosophy again. Of course, each fork can only be held by one philosopher at a time and so when a philosopher tries to grab a fork, it may or may not be available.

    a. Assume one philosopher is an altruist. That is, in the case that every philosopher holds his left fork, this philosopher will return his fork to the table. Implement in SMV a system that encodes the philosopher dining problem.

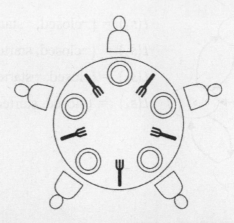

FIGURE 7.17
Five dining philosophers.

    b. Verify the correctness of the system by specifying and checking the following properties:

       i. Two neighboring philosophers never eat at the same time.

      ii. No more than two philosophers can eat at the same time.

     iii. Every philosopher can eat infinitely often.

9. Consider the classic missionary and cannibal problem. Three missionaries and three cannibals want to cross a river but there is only one boat available that holds up to two passengers at a time. If the cannibals ever outnumber the missionaries on either bank, the missionaries will be eaten. The boat cannot cross the river by itself with no people on board. The problem consists of finding a strategy that will enable them to cross the river safely. Implement in SMV a system that encodes the above problem, and prove with NuSMV that there exists a solution to the problem. Notice that you need to show how the missionaries and cannibals cross the river step-by-step.

8

Petri Nets

The classic Petri net was invented by Carl Adam Petri in 1962. Since then, a lot of research has been conducted to fully explore the utility of Petri nets. Until 1985, the Petri net was mainly used by theoreticians as an interest subject of mathematics. However, their practical use has increased due to the introduction of high-level Petri nets. Adoption of Petri nets is facilitated by the availability of tools for diagramming, defining, and using them.

8.1 Petri Nets

A Petri net is a bipartite directed graph composed of places and transitions. An example of a Petri net is illustrated in Figure 8.1. The circles, squares, dots, and arrows all have specific meanings that are used to model a system. A *place* is denoted by a circle and a *transition* is denoted by a square. The connections, called *arcs*, are directed and are situated between a place and a transition with the direction indicated by the arrowhead. *Tokens* are dynamic objects used to track information and appear as solid dots within the circle of a place. The labels for places and transitions can appear in or around the item with a location that is established to facilitate understandability.

The state of a Petri net, called the *marking*, is determined by the distribution of tokens over the places. A marking is obvious from a figure. In the example in Figure 8.1, we can say that p_1 has 1 token, p_2 has 2 tokens, p_3 has 0 tokens, and p_4 has 0 tokens. A more convenient notation would show that marking as (1, 2, 0, 0) where the first entry is for p_1, the second for p_2, and so on.

In Figure 8.1, we can say that transition t_1 has one input place, p_1, and two output arcs connecting to places, p_2 and p_3. Transition t_3 has two input places, p_2 and p_3, and two output places, p_3 and p_4. Place p_3 is both an input place and an output place for t_3. Transition t_2 has two input places, p_2 and p_4, and no output places.

Table 8.1 shows typical interpretations of transitions and places. Input places can serve as preconditions, input data, input signals, required resources,

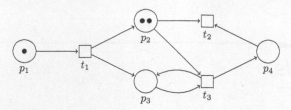

FIGURE 8.1
Example Petri net.

TABLE 8.1

Typical Interpretations of Transitions and Places

Input Places	Transitions	Output Places
Precondition	Event	Postcondition
Input Data	Computation Step	Output Data
Input Signal	Signal Processor	Output Signals
Required Resources	Task or Job	Released Resources
Conditions	Logic Clause	Conclusion
Buffers	Process	Buffers

conditions, or buffers. The output places can serve as post conditions, output data, output signals, resources released, conclusions, or buffers. The transitions can describe events, computation steps, signal processes, tasks, jobs, logic clauses, or even processors.

Transitions are active components, while places and tokens are passive. Place is static, that is, once the Petri net is defined the places do not change. Tokens are dynamic, since they are created and consumed as a result of a transition. It is by firing a transition that the marking of a Petri net is changed.

A transition is *enabled* (able to be fired) if each of its input places contains tokens. In Figure 8.1, transitions t_1 and t_2 are enabled, while transition t_3 is not. Transition t_1 is enabled because the only input place p_1 contains a token. Transition t_2 is enabled because there is a token in p_2. Transition t_3 is not enabled because input place p_3 does not have a token even though the other input place, p_2, has a token. For transition t_3 to be enabled there would have to be tokens in places p_2 and p_3.

A transition that is enabled can *fire*. When a transition fires, it consumes a token from each input place (removes it) and creates a token in each output place. Firing is atomic in the sense that all of the consumption and creation of tokens takes place at one time (it is assumed to occur instantaneously).

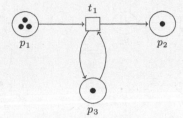

FIGURE 8.2
Example 8.1.

Example 8.1

Consider the Petri net in Figure 8.2. It has three places, p_1, p_2, and p_3, with a single transition, t_1. The initial marking is (3, 1, 1). Transition t_1 has two input places, p_1 and p_3. It has two output places, p_2 and p_3. In the initial state, transition t_1 is enabled since it has tokens in each of its input places. What happens when transition t_1 fires?

When transition t_1 fires, the following actions all occur simultaneously:

- A token is consumed from p_1.
- A token is consumed from p_3.
- A token is created in p_2.
- A token is created in p_3.

The result leaves the Petri net in a new state described with the marking (2, 2, 1) as shown in Figure 8.3. It is important to understand that firing transition t_1 should not be interpreted as a token moving from p_1 to p_2 and the token in p_3 moving and moving back. Tokens are consumed and created—not moved! The interpretation of tokens moving about implies something that isn't actually happening in the real system.

As seen in the right side of Figure 8.3, transition t_1 is still enabled because tokens remain in places p_1 and p_3. What happens if transition t_1

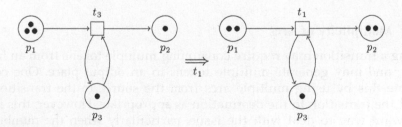

FIGURE 8.3
Firing t_1 (first time).

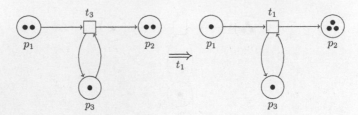

FIGURE 8.4
Firing t_1 (second time).

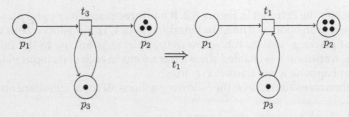

FIGURE 8.5
Firing t_1 (third time).

is fired again? The same set of actions that occurred before are repeated: tokens are consumed in p_1 and p_3 and tokens are created in p_2 and p_3. The result leaves the Petri net in a state described by marking (1, 3, 1). This is shown in Figure 8.4.

Transition t_1 is still enabled because tokens remain in places p_1 and p_3. What happens if transition t_1 is fired again? The same actions that occurred before: tokens are consumed in p_1 and p_3 and tokens are created in p_2 and p_3. The result leaves the Petri net in a state described by marking (0, 4, 1). This is shown in Figure 8.5. At this point, transition t_1 is no longer enabled since place p_1 does not have a token in it.

8.1.1 Multiplicity of Arcs

Firing a transition may require consuming multiple tokens from an input place and may generate multiple tokens to an output place. One could denote this by using multiple arcs from the source to the transition or from the transition to the destination as appropriate. However, this is an awkward way to deal with the issue, particularly when the number of tokens is large.

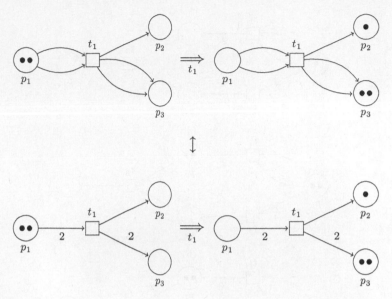

FIGURE 8.6
Multiplicity of arcs.

Instead, we can incorporate multiplex arcs in which the graphical notation is extended to allow a number to be associated with the arc denoting how many tokens are destroyed or created, as appropriate. Figure 8.6 shows the equivalence between using multiple arcs from place to transition and transition to place, and using multiplex arcs.

Example 8.2

Figure 8.7 illustrates a Petri net with multiplex arcs and the markings after t_1, t_2, and t_3 have fired. Figure 8.7a identifies the initial marking, $(1, 0, 0, 0)$. Only a single transition, t_1, is enabled under this initial marking.

Firing t_1 consumes the token in p_1 and creates 2 tokens in p_1 and 2 tokens in p_2 resulting in the marking $(2, 2, 0, 0)$. Figure 8.7b shows the result of firing t_1. At this point, transitions t_1 and t_2 are both enabled. The transition t_3 is not yet enabled since there isn't a token in p_3.

Firing t_2 consumes the 2 tokens in p_1 and creates a token in p_3 as shown in Figure 8.7c. The new marking is $(0, 2, 1, 0)$. At this point, the only transition that is enabled is t_3 since p_1 no longer has any tokens in it.

Firing t_3 consumes the 2 tokens in p_2 and the single token in p_3 while producing a single token in p_4 as shown in Figure 8.7d. The marking is now $(0, 0, 0, 1)$ and no transitions are enabled.

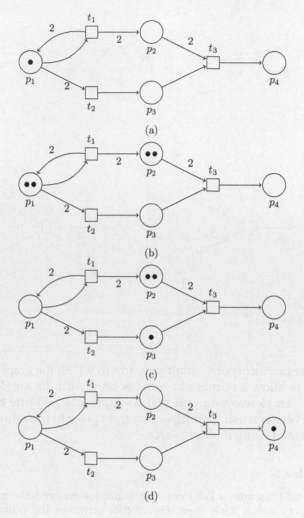

FIGURE 8.7
Petri net transitions with multiplex arcs: (a) initial marking, (b) firing t_1, (c) firing t_2, and (d) firing t_3.

8.2 Common Petri Net Structures and Substructures

Some arrangements of states and transitions within a Petri net tend to describe a basic kind of behavior for the system being modeled. The value of understanding the common structures is that it allows one to comprehend the overall behavior of a system and to more easily use a Petri net to specify a behavior.

8.2.1 Sequential Execution

One of the most common specification activities is to describe the execution of a sequence of operations. The sequence is imposed by tokens that are consumed in one step and created in the next with simple transitions linking the input place and the output place. A Petri net for specifying sequential execution is shown in Figure 8.8.

When a token is placed in p_1, only transition t_1 is enabled. Firing t_1 consumes the token in p_1 and creates a token in p_2. The absence of a token in place p_1 prevents t_1 from being enabled. However, the token in p_2 allows t_2 to be enabled. Firing t_2 removes the token from p_2 and creates a token in p_3. Transition t_3 is now the only transition that is enabled. When it fires, it removes the token in p_3 and places it on whatever place is an output for t_3. This enforces a firing sequence of t_1, t_2, t_3, and so on.

8.2.2 Concurrent Execution

Often one has to deal with situations where the program has to do multiple things all at the same. For example, it might need to load a file while listening for mouse event to cancel loading the file. This kind of behavior requires a program to pursue multiple paths of execution that are usually, but not always, independent of each other. The Petri net for describing this kind of situation appear in Figure 8.9.

When a token is placed in p_1, the transition t_1 is enabled. Firing t_1 consumes the token at p_1 and creates tokens in p_2 and p_4. At that point, both transitions, t_2 and t_2, are enabled. Each can fire independently of the other. What this produces is two different paths of sequential execution that can operate independently of each other.

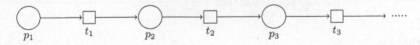

FIGURE 8.8
Sequential execution.

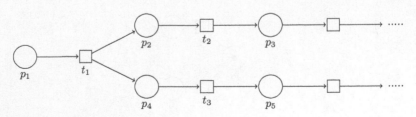

FIGURE 8.9
Concurrent execution.

8.2.3 Synchronization

In Figure 8.9, transition t_1 splits execution into two independent sequences. Synchronization is the opposite of that splitting. It is taking independent sequences and forcing them to reach specific states before continuing on with execution. This is shown in Figure 8.10.

In Figure 8.10, there are three concurrent sequential paths of execution. Each path of execution can proceed independently of the others until the path reaches of input places to transition t_7. This transition can only fire when places p_4, p_5, and p_6 all have a token within them. What this does is force all of the concurrent threads to come to completion before execution can continue.

8.2.4 Nondeterminism

A problem that can creep into a specification is that of nondeterminism. This is a situation in which there are two possible paths of execution but nothing to uniquely determine which path should be followed. Such a situation can be documented using Petri nets as illustrated in Figure 8.11.

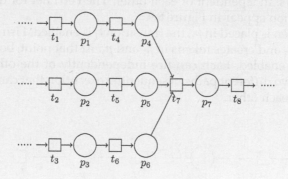

FIGURE 8.10
Synchronization.

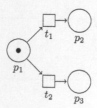

FIGURE 8.11
Nondeterminism.

What is shown in Figure 8.11 is a case where two transitions are enabled, t_1 and t_2, as a result of the same input place holding a token. At this point, there is nothing dictating that t_1 or t_2 should be the next transition to fire. If either transition is fired, the other transition is disabled. Putting more tokens in place p_1 doesn't resolve the indeterminate situation. With 2 tokens it is possible for t_1 to fire twice, t_2 to fire twice, t_1 to fire followed by t_2 firing, or t_2 to fire followed by t_1 firing.

8.2.5 Loop

One of the most common control structures is a loop. A Petri net with loop is shown in Figure 8.12.

In a Petri net the loop basically has tokens being destroyed and created in a cyclic fashion. This figure shows a loop without a control structure that would cause it to come to an end. Unfortunately, there really isn't a way to dynamically test for eminent conditions on loops within a Petri net without incorporating nondeterminism or a control, such as shown later.

8.2.6 Source

A source is a transition that has no input place but does have an output place. This is shown in Figure 8.13. The absence of an input place allows the transition to be enabled at all times.

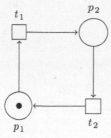

FIGURE 8.12
Loop.

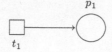

FIGURE 8.13
Source.

Without the need to consume a token from a source place, transition t_1 is enabled at all times and thus can fire, producing a token in place p_1. This can be used to model an infinite source of resources entering into the system.

8.2.7 Consumer

The opposite of a source is a consumer. A consumer is a transition that has an input place with no output places, such as shown in Figure 8.14. This consumes the tokens in the input place for the transition.

8.2.8 Control

Modeling some kind of fixed resource can be accomplished rather easily. In this case, a control place is initialized with the available resources. That place can be incorporated into the Petri net as a limiter. This is illustrated in Figure 8.15. It can also be used as a break point for the loop. In other words, the loop repeats until the control is empty.

Place p_1 represents a consumable resource. While t_1 and t_2 essentially complete a loop through places p_2 and p_3. The key thing to notice is that t_1 can only fire three times because after the third time all of the tokens in enabling place p_1 will have been consumed.

FIGURE 8.14
Consumer.

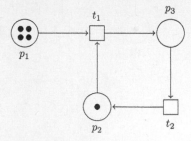

FIGURE 8.15
Control.

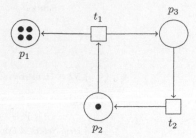

FIGURE 8.16
Accumulator.

8.2.9 Accumulator

The opposite of a control is an accumulator. Rather than a place that holds tokens which are to be consumed, an accumulator is a place where tokens are created and never consumed, as shown in Figure 8.16. In this case, place p_1 accumulates a token each time t_1 is fired. Since nothing is consuming the tokens created in place p_1, it effectively holds a count of how many times t_1 has been fired. Hence the name accumulator.

Example 8.3

Figure 8.17 shows a Petri net representing a general announcement program that displays news items on a monitor and speaks the news item aloud over a loudspeaker. It is easy to analyze this program by recognizing the various structures and substructures within it.

At the top of the Petri net is a transition, t_1, that serves as a source capable of producing an infinite number of articles. The places *News Item Received* and *Item Selected For Processing* are part of a sequential execution with a source capable of generating as many news items as necessary, which represented by the transition t_1. Transition t_2 turns the sequential execution into a pair of concurrent executing threads. These threads are synchronized as a result of transition t_6. Transition t_7 completes the announcement process by consuming the token representing the news item.

It is fairly simple to see how to modify the behavior of this system to transform the basic functionality, for example, replacing a scroll monitor display with a text message. In this case, one would remove the left-hand thread of the concurrent section and replace it with the sequential Petri net for sending it as a text message. Alternatively, one could add that functionality by adding another concurrent sequence by splitting it into three threads rather than two. This would have no effect on the overall flow of the program.

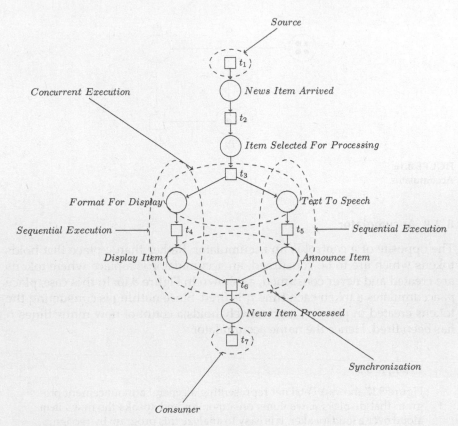

FIGURE 8.17
Automated public announcement system.

8.3 Reduction Rules

A severe limitation to Petri net modeling is in the large size of resulting nets, especially when dealing with real systems at a detailed level. Therefore, it is helpful to reduce the system model to a simple one, while preserving the system properties to be analyzed. We can employ some simple transformations, called reduction rules, that can be used for analyzing a complex Petri net by simplifying it to something that is easily understood. The reduction rules can actually be used in both directions: simplifying a Petri net or adding details to a Petri net.

8.3.1 Fusion of Series Places

Suppose there is a situation in which there are two places and a single transition that connects them in series. A Petri net for this case is

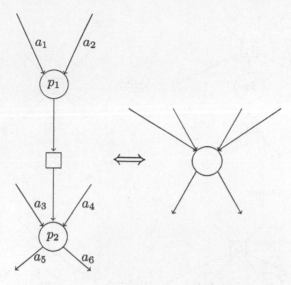

FIGURE 8.18
Fusion of series places.

illustrated in Figure 8.18. Labeling the incoming arcs as, a_1, a_2, a_3, and a_4, one can see that any token which arrives via any of these arcs will ultimately traverse to p_2. Tokens that are created as a result of a_1 and a_2 will reside in place, p_1, and enable transition t_1. When transition t_1 fires, the token will move to place p_2. In this configuration, there is no other transition out of place p_1. Under these conditions, the two places can be combined into one place. Thus there is an equivalence in the two Petri nets shown in the figure.

8.3.2 Fusion of Series Transitions

In this case, there are two transitions separated by a single place. When transition t_1 fires, a token is created in place p_1. This enables transition t_2, which will consume the token in p_1. Tokens will be generated in the places of the other outgoing arcs of t_1 and t_2. The result is identical markings created in the case shown to the right of Figure 8.19.

8.3.3 Fusion of Parallel Places

In Figure 8.20, transition t_1 creates a token in p_1 and p_2. Transition t_2 is the only transition with p_1 and p_2 as sources. Firing t_2 consumes the tokens in p_1 and p_2, generating the tokens on the outgoing arc of t_2. Thus the left and right sides of Figure 8.20 are equivalent.

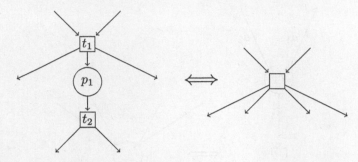

FIGURE 8.19
Fusion of series transitions.

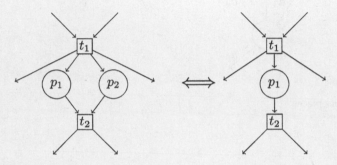

FIGURE 8.20
Fusion of parallel places.

8.3.4 Fusion of Parallel Transitions

Figure 8.21 shows two places, p_1 and p_2, that are connected by two transitions, t_1 and t_2. Both t_1 and t_2 consume a token from p_1 and produce a token in p_2. Regardless of which transition fires, the effect is to consume a token from p_1 and produce one in p_2, thus the transitions are redundant in terms of the behavior of the Petri net. The transitions can be fused into a single transition as shown to the right of the double-headed arrow.

8.3.5 Elimination of Self-Loop Places

The self-loop place shown in Figure 8.22 does not add anything to the Petri net. That place will never change its marking regardless of how many times the transition is fired.

8.3.6 Elimination of Self-Loop Transitions

A self-loop on a transition, shown on the left of Figure 8.23, does not change the marking upon firing. Because the marking is unchanged, it can be removed as shown on the right of Figure 8.23.

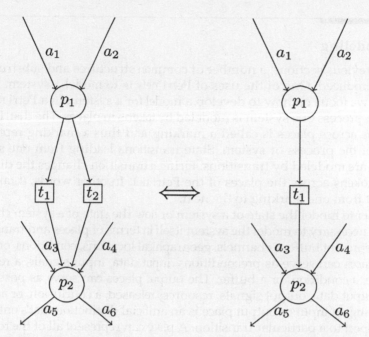

FIGURE 8.21
Fusion of parallel transitions.

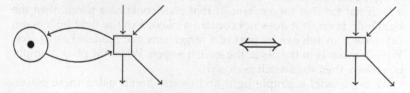

FIGURE 8.22
Elimination of self-loop places.

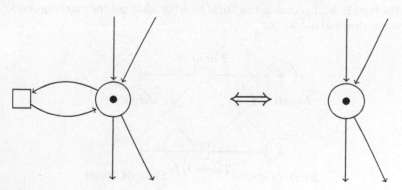

FIGURE 8.23
Elimination of self-loop transitions.

8.4 Modeling

In the previous sections, a number of common structures and substructures were introduced. One of the uses of Petri nets is to model a system. In this section, we focus on how to develop a model for a system. In a Petri net, the state of a process or a system is modeled by tokens in places. The distribution of tokens across places is called a marking, and thus a marking represents a state of the process or system. State transitions leading from one state to another are modeled by transitions. Firing a transition changes the distribution of tokens across the places of the Petri net. In other words, it takes the Petri net from one marking to the next.

In order to model the state of a system or how the state of a system changes, it is first necessary to model the system itself in terms of places and transitions. Places represent buffers, channels, geographical locations, conditions, or states. Input places can serve as preconditions, input data, input signals, a required resource, a condition, or a buffer. The output places can serve as post conditions, output data, output signals, resources released, a conclusion, or a buffer. Place being an input or output place is an artificial distinction that's only valid with respect to a particular transition. A place can represent all of the roles.

Example 8.4

A light bulb that is part of a larger item can be modeled as a place in a Petri net for that larger item. If that place contains a token, then the light bulb is on. If it does not contain a token then the light bulb is off. Likewise a switch can be part of a larger item and modeled as a place. When a token is in the place, the switch is open. When the place is empty of a token, then the switch is closed.

We can model a simple light and switch circuit using these places. The two transitions are: TurnOn and TurnOff. The result is shown in Figure 8.24. Initially, the switch is open and the light is off. Executing the TurnOn action changes the markings such that the switch is closed and the light is on. Executing the TurnOff action changes the markings back to the original marking.

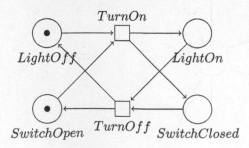

FIGURE 8.24
Petri net for light bulb.

Transitions represent events, transformations, or transportations. In a lot of modeling activities, it is the transitions that are the more interesting elements. The transitions do things. One can almost view the places as points between the executions of actions performed by the system. A place is a point where the system can decide on what is the next transition that should be made.

In modeling a system, one can come up with transitions that encapsulate a number of activities. Further elicitation of what's involved in a transition can have the effect of transforming a single transition into a chain of sequentially executed transitions. This is a functional decomposition in which some function is decomposed into finer-grained functions. This is shown in Example 8.5.

Example 8.5

There is a party and it is desired to extend a written invitation for particular friend to attend. Thus we can have two places, one representing no invitation and one representing an invitation has been extended. There's a simple transition, extend invitation. Such a model is shown in leftmost figure of Figure 8.25.

This model is overly simple. It's much like that cartoon with the step which is labeled, "Then a miracle occurs here." We know something is supposed to happen, but not what.

Details for the model can be introduced by breaking the "extend invitation" transition into a sequence of transitions and places. The transition "extend invitation" can be broken down into "write invitation" and "deliver invitation." The newly introduced intermediate place represents that an invitation has been created. This revised Petri net is illustrated in the second figure from left in Figure 8.25.

The transition for "write invitation" can be broken down into additional transitions. We know that in order to write an invitation, we need paper, pen, and envelope. It is necessary then to sit down with pen and paper to compose the invitation. Thus "write invitation" can be decomposed into two transitions "gather materials" and "write invitation," with a newly created place representing materials gathered. This revised Petri net is illustrated in the third figure from left in Figure 8.25.

Further decomposition is possible, but one in particular is of interest here, that being "gather materials." As stated, it was determined that before writing a letter one should gather together paper, pen, and envelop. There is a modeling choice that can be made here concerning how one gathers material.

One can model it such that one gets the paper, then gets the pen, and then gets the envelope in sequence. The problem here is that the order in which the materials are gathered is fully specified. According to the Petri net, one can't get the envelope, then the pen, and finally the paper.

An alternative is to break the "gather materials" transition into concurrent sequences, which allows for the collection of the items in any order one wants, but prevents sitting down to compose the invitation until after all items are collected. This is shown at the rightmost side of

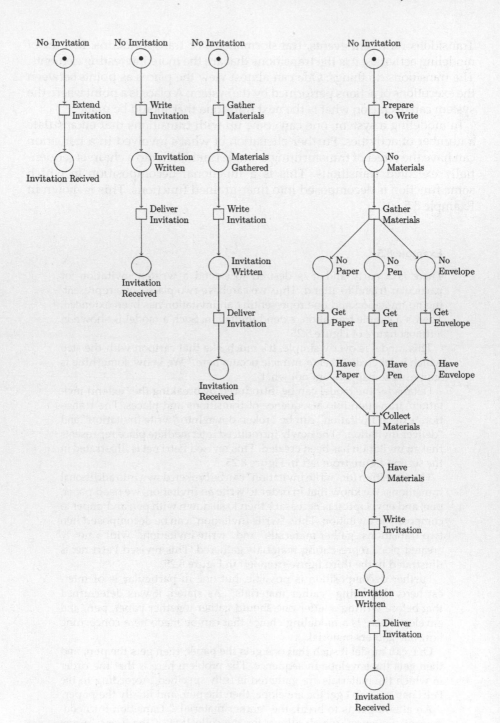

FIGURE 8.25
Write an invitation letter.

Figure 8.25. This introduces a number of intermediate states: need pen, pen acquired, need paper, paper acquired, need envelope, and enveloper acquired.

Finally, with the situation being modeled, we would have an initial marking with a single token in the place representing "no invitation."

Example 8.6

Petri nets can be used to model everyday systems like traffic lights. A traffic light has three states: red light, green light, and yellow light. Only one state can be active at a time. The transitions are: red → green (rg), green → yellow (gy), and yellow → red (yr). The token identifies the current state of the traffic light. A Petri net for a simple traffic light (northbound–southbound traffic) is shown in Figure 8.26.

Of course, traffic control does not depend on a single light in one direction. There would also be an east–west road crossing it. To capture both east–west and north–south traffic control, one would need two Petri nets, one for each direction. Such a configuration is shown in Figure 8.27.

The problem is that these two uncoupled Petri nets could be green in both directions leading to risky times and accidents. In order to make these lights safe, a resource node needs to be added that is consumed by one direction or the other while blocking the direction that doesn't consume the resource. A Petri net for this case is illustrated in Figure 8.28.

The safe traffic light has one minor problem, one direction can cycle through the lights while blocking the other direction from moving off red. That is, transition rg_1 can fire multiple times without allowing transitions rg_2 from firing. To correct this, an additional place is added so that the resource is transferred from the light in one direction to the light in the other. This is shown in Figure 8.29.

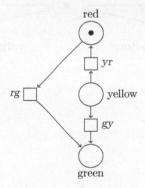

North South Light

FIGURE 8.26
Simple traffic light.

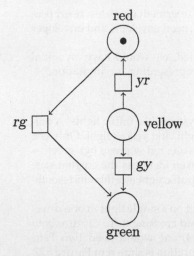
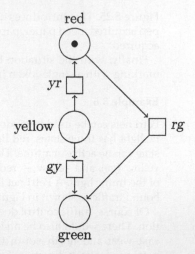

North South Light   **East West Light**

FIGURE 8.27
Pair of traffic lights.

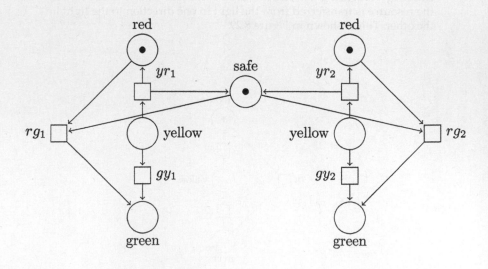

North South Light   **East West Light**

FIGURE 8.28
Safe traffic light.

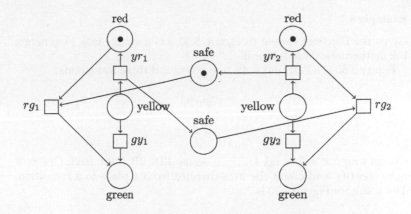

North South Light   East West Light

FIGURE 8.29
Safe and fair traffic light.

8.5 Mathematical Description of Petri Nets

A graphical depiction of a Petri net allows one to understand it structurally and intuitively. However, a picture doesn't help answer a number of questions that are of interest when specifying a system, including such questions as

- Can the system reach a specific state from the current state?
- What states of the system are accessible?
- Is the system alive?

To answer these questions, one must adopt a mathematical model of a Petri net. A Petri net N is a tuple $N = \{P, T, I, O, M_0\}$, where

- P is a finite set of places, graphically represented by circles.
- T is a finite set of transitions, graphically represented by boxes.
- $I: P \times T \rightarrow (\{0, 1, 2, ...\})$ is the pre-incidence function representing input arcs.
- $O: T \times P \rightarrow (\{0, 1, 2, ...\})$ is the post-incidence function representing output arcs.
- $M_0: P \rightarrow N$ is the initial marking representing the initial distribution of tokens.

Places P and transitions T are disjoint ($P \cap T = \varnothing$).

Example 8.7

Given the Petri net shown in Figure 8.30, it is a simple task to generate the mathematical model for it.

Figure 8.30 is a Petri net with two places and three transitions:

$$P = \{p_1, p_2\}$$

$$T = \{t_1, t_2, t_3\}$$

Computing $I: P \times T \rightarrow (\{0, 1, 2, ...\})$ seems difficult, but it isn't. One task is to identify and count the arcs directed from a place to a transition. The result for Figure 8.30 is

$$I(p_1, t_1) = 1 \quad I(p_1, t_2) = 2 \quad I(p_1, t_3) = 0$$

$$I(p_2, t_1) = 1 \quad I(p_2, t_2) = 0 \quad I(p_2, t_3) = 2$$

Likewise, $O: T \times P \rightarrow (\{0, 1, 2, ...\})$ is determined by identifying and counting the arc directed from a transition to a place, with the result for Figure 8.30 being

$$O(t_1, p_1) = 1 \quad O(t_1, p_2) = 0$$

$$O(t_2, p_1) = 0 \quad O(t_2, p_2) = 1$$

$$O(t_3, p_1) = 0 \quad O(t_3, p_2) = 1$$

Since there are 3 tokens in p_1 and 2 tokens in p_2, the initial marking for the example is

$$M_0 = (3, 2)$$

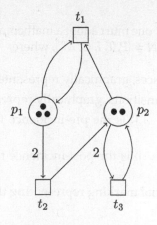

FIGURE 8.30
Example Petri net for mathematical modeling.

An alternative way to represent the transition functions is to suppress the place as an argument and using place in the output of the functions to correlate place and count of directed arcs into or out of the place. Thus,

$$I(p_1, t_1) = 1, I(p_2, t_1) = 1 \rightarrow I(t_1) = (1, 1)$$

$$O(t_1, p_1) = 1, O(t_1, p_2) = 0 \rightarrow O(t_1) = (1, 0)$$

Doing this for all of the transitions yields

$$I(t_1) = (1, 1) \quad I(t_2) = (2, 0) \quad I(t_3) = (0, 2)$$

$$O(t_1) = (1, 0) \quad O(t_2) = (0, 1) \quad O(t_3) = (0, 1)$$

What is nice about this form of the pre-incidence and post-incidence functions is how easy it is to determine if a transition is enabled in a current marking. In particular whether each position in the current marking is greater than or equal to the corresponding position in the pre-incidence function for the transition.

Given the initial marking, $M_0 = (3, 2)$, then it can be observed that

$$M_0 > I(t_1)$$

$$M_0 > I(t_2)$$

$$M_0 > I(t_3)$$

Thus, all three transitions are enabled from the initial marking.

A transition t is enabled at marking M_i if and only if

$$M_i \geq I(t)$$

Let $E(M_i)$ be the set of all transitions enabled at M_i, then $t \in E(M_i)$. Suppose that the firing of t takes the Petri net from M_i to M_j, then we can say

$$M_j = M_i - I(t) + O(t)$$

The $I(t)$ is subtracted because it represents the tokens that are consumed in M_i. The $O(t)$ is added because it represents the tokens created. The effect of making the transition can be denoted by $M_j [t > M_i$.

Example 8.8

Given the Petri Net of the previous example with an initial marking illustrated in Figure 8.30. We can say the following:

$$P = \{p_1, p_2\}$$

$$T = \{t_1, t_2, t_3\}$$

$$I(t_1) = (1, 1), \quad I(t_2) = (2, 0), \quad I(t_3) = (0, 2)$$

$$O(t_1) = (1, 0), \quad O(t_2) = (0, 1), \quad O(t_3) = (0, 1)$$

$$M_0 = (3, 2)$$

The transitions that are enabled within the Marking M_0 are

$$E(M_0) = \{t_1, t_2, t_3\}$$

If transition t_1 is fired, then

$$M_0 [t_1 > M_1$$

where $M_1 = (3, 1)$ since

$$(3, 2) - (1,1) + (1, 0) \rightarrow (3 - 1 + 1, 2 - 1 + 0) \rightarrow (3, 1)$$

This is very simple to implement in a program.

8.6 Petri Net Behavior

Given the mathematical description of a Petri Net, it is now possible to compute how a Petri net can behave. Some things that we will want to know include: What states are reachable? Is the Petri net bounded? Is there a deadlock? Are there sinks that accumulate tokens or syphons that generate tokens?

8.6.1 Reachability

Reachability is a fundamental basis for studying dynamic properties of any system. A marking M_n is said to be reachable from a marking M_0 if there exists a sequence of firings that transform $M_0 - M_n$. A marking M_j is said to be immediately reachable from M_i if firing an enabled transition in M_i results in M_j. The set of all markings reachable from marking M is denoted by $R(M)$.

8.6.1.1 The Reachability Tree

Given a Petri net and an initial marking, it is sometimes desirable to know all of the states that can be reached by firing transitions. The *reachability tree* for a Petri net N is constructed by the following algorithm:

1. Label the initial marking M_0 as the root and tag it "new."
2. For every new marking M:
    a. If M is identical to a marking already appeared in the tree, then tag M "old" and go to another new marking.

b. If no transitions are enabled at M, tag M "dead end" and go to another new marking.

c. While there exist enabled transitions at M, do the following for each enabled transition t at M:

   i. Obtain the marking M' that results from firing t at M.

   ii. Introduce M' as a node, draw an arc with label t from M to M', and tag M' "new."

Example 8.9

Consider the Petri Net in Figure 8.31. What are all of the possible markings? The initial marking for this Petri net is

$$M_0 = (2, 0, 0, 0) \quad \text{new}$$

$E(M_0) = \{t_1\}$, which says that the only enabled transition is t_1 which upon firing produces, M_1:

$$M_0 [t_1 > M_1 = (0, 2, 1, 0) \quad \text{new}$$

$E(M_1) = \{t_2, t_3\}$ which gives two possible firings t_2 and t_3. The order of firing is important.

$$M_1 [t_2 > M_2 = (0, 1, 1, 1), \quad \text{new}$$

$$M_1 [t_3 > M_3 = (0, 2, 0, 1), \quad \text{new}$$

There are now two branches to follow, one starting with M_2 and the other with M_3. $E(M_2) = \{t_2, t_3\}$, which means that there are again two possible firings:

$$M_2 [t_3 > M_4 = (0, 1, 0, 2) \quad \text{new}$$

$$M_2 [t_2 > M_5 = (0, 0, 1, 2) \quad \text{new}$$

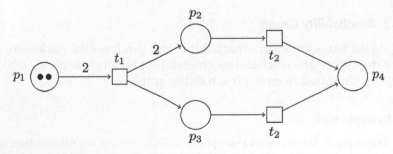

FIGURE 8.31
Example for reachability.

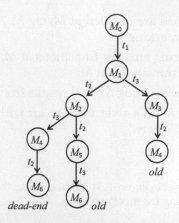

FIGURE 8.32
Reachability tree for Example 8.9.

$E(M_4) = \{t_2\}$, meaning there is only a single transition which is enabled. The result of firing t_2 is

$$M_4[t_2 > M_6 = (0, 0, 0, 3) \qquad \text{dead end}$$

This transition is marked dead since $E(M_6) = \{\}$. Returning to the last branch point that produced M_4 and M_5, we can follow the second branch, $E(M_5) = \{t_3\}$.

$$M_5[t_3 > M_7 = M_6 = (0, 0, 0, 3) \qquad \text{old}$$

This is marked old since M_7 has the same markings as M_6. Now the algorithm returns to the second branch with M_3. $E(M_3) = \{t_2\}$ which allows:

$$M_3[t_2 > M_8 = M_4 = (0, 1, 0, 2) \qquad \text{old}$$

At this point there are no further transitions that need to be considered. This sequence of transitions and markings can be illustrated as a tree (Figure 8.32).

8.6.1.2 Reachability Graph

Merging the same nodes in a reachability tree produces the *reachability graph*. One of the values of a reachability graph is that is can show cycles, which are less easily identified from the reachability graph.

Example 8.10

Returning to the previous example for calculating the reachability tree, it is simple to construct the reachability graph. The two states marked old are connected to the states that they duplicate. This is shown in Figure 8.33.

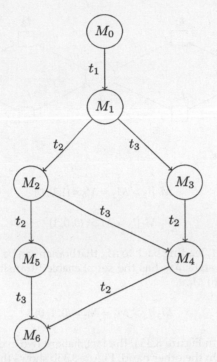

FIGURE 8.33
Reachability graph for Example 8.9.

Example 8.11

Consider the simple graph shown in Figure 8.34. This graph has a large exterior cycle encompassing two smaller internal cycles. While this example is pretty obvious in that the two cycles exist, this may not be the case for more complex Petri nets. Still, the construction of a reachability graph identifies loops in a more direct manner.

The reachability tree and graph are constructed following the same process as in the previous example. The initial marking is

$$M_0 = (1, 0, 0)$$

The set of enabled transitions is given by $E(M_0) = \{t_1\}$. There is only one transition that is enabled from the initial marking. The result of firing it is

$$M_0 [t_1 > M_1 = (0, 1, 0)$$

The set of enabled transitions is $E(M_1) = \{t_2, t_3\}$. There are two transitions that are enabled, which when firing give

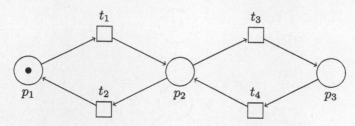

FIGURE 8.34
Petri net with two cycles.

$$M_1[t_2 > M_2 = M_0 = (1, 0, 0)$$

$$M_1[t_3 > M_3 = (0, 0, 1)$$

With transition t_2 leading back to M_0, that branch can be dismissed as old. The remaining state, M_3, has the set of enabled transitions $E(M_3) = \{t_4\}$. Firing t_2, leads to M_1,

$$M_3[t_4 > M_4 = M_1 = (0, 1, 0)$$

As can be seen in Figure 8.35a, the fact that are two cycles isn't immediately obvious. On the other hand, Figure 8.35b shows the cycles in a very obvious manner.

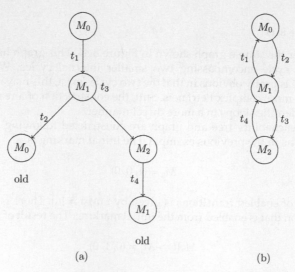

FIGURE 8.35
Example 8.11: (a) Reachability tree. (b) Reachability graph.

8.6.2 Boundedness

In a Petri net, places are often used to represent information storage areas in communication and computer systems, product and tool storage areas in manufacturing systems, and so on. It is important to be able to determine whether proposed control strategies prevent overflows from these storage areas. The Petri net property that helps to identify the existence of overflows in the modeled system is the concept of boundedness.

A place p is said to be k-bounded if the number of tokens in p is always less than or equal to k for every marking M reachable from the initial marking M_0, that is, $M \in R(M_0)$. Since it is not possible to "borrow" tokens that don't exist, it follows that k must be a nonnegative integer number. A place is safe if it is 1-bounded.

Example 8.12

Figure 8.34 shows a Petri net where p_1 is 1-bounded. To understand why, all one has to do is examine all of the markings reachable from (M_0), which is given by

$$R(M_0) = \{(1,0,0), (0,1,0), (0,0,1)\}$$

The greatest number of tokens for place p_1 appearing in the reachability set is 1. The same is true for places, p_2 and p_3.

A Petri net $N = (P, T, I, O, M_0)$ is k-bounded (safe) if each place in P is k-bounded (safe). Returning to the Petri net in Figure 8.34, it is 1-bounded since all places in it are 1-bounded. A Petri net is unbounded if k is infinitely large.

8.6.3 Liveness

In computer science, deadlock refers to a specific condition in which the processes of a group are each waiting for another process to perform some action. The result is that none of the processes in that group are able to perform an action. Deadlock often results from resource sharing and synchronization of parallel processes. Deadlock has been studied extensively in the context of computer operating systems.

Liveness is a concept closely related to deadlock. A Petri net modeling a deadlock-free system must be live. Liveness implies the absence of total or partial deadlock and is often required for well-designed systems. But the reverse is not true. We can have a system that avoids deadlock, but is not live.

A transition t is said to be live if it can always be made enabled starting from any reachable marking, that is,

$$\forall M \in R(M_0), \exists M' \in R(M) \text{ such that } M'[t>$$

This implies that for any reachable marking M, it is ultimately possible to fire any transition in the net by progressing through some firing sequence. The "complete" liveness requirement, however, might be too strict to represent some real systems or scenarios that exhibit deadlock-free behavior.

For instance, the initialization of a system can be modeled by a transition (or a set of transitions) that fires a finite number of times. After initialization, the system may exhibit a deadlock-free behavior, although the Petri net representing this system is no longer live as specified above. For this reason, different levels of liveness are defined.

Denote by $L(M_0)$ the set of all possible firing sequences starting from M_0. A transition t in a Petri net is said to be

L0-live (or dead): If there is no firing sequence in $L(M_0)$ in which t can fire.

L1-live (potentially fireable): If t can be fired *at least once* in *some* firing sequence in $L(M_0)$.

L2-live: If t can be fired *at least k times* in *some* firing sequence in $L(M_0)$ given any positive integer k.

L3-live: If t can be fired *infinitely* often in *some* firing sequence in $L(M_0)$.

L4-live (or live): If t is L1-live (potentially fireable) in *every* marking in $R(M_0)$.

A Petri net is said to be live if all transitions are live. A Petri net is said to be quasi-live if all transitions are quasi-live. A marking M is said to be a deadlock or dead marking if no transition is enabled at M. A Petri net is said to be deadlock-free if it does not contain any deadlock in any reachable marking from the initial marking.

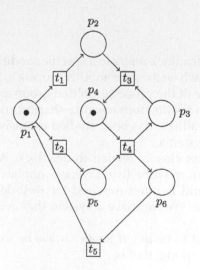

FIGURE 8.36
A safe, nonlive Petri net.

Example 8.13

Figure 8.36 show a safe, nonlive Petri net. But it is strictly L1-live.

$$I(t_1) = (1,0,0,0,0,0) \qquad O(t_1) = (0,1,0,0,0,0)$$

$$I(t_2) = (1,0,0,0,0,0) \qquad O(t_2) = (0,0,0,0,1,0)$$

$$I(t_3) = (0,1,1,0,0,0) \qquad O(t_3) = (0,0,0,1,0,0)$$

$$I(t_4) = (0,0,0,1,1,0) \qquad O(t_4) = (0,0,1,0,0,1)$$

$$I(t_5) = (0,0,0,0,0,1) \qquad O(t_5) = (1,0,0,0,0,0)$$

$$L(M_0) =$$

$$M_0 = (1,0,0,1,0,0) \qquad \{t_1,t_2\}$$

$$M_1 = M_0[t_1> = (0,1,0,1,0,0) \qquad \{\}$$

$$M_2 = M_0[t_2> = (0,0,0,1,1,0) \qquad \{t_4\}$$

$$M_3 = M_2[t_4> = (0,0,1,0,0,1) \qquad \{t_5\}$$

$$M_4 = M_3[t_5> = (1,0,1,0,0,0) \qquad \{t_1,t_2\}$$

$$M_5 = M_4[t_1> = (0,1,1,0,0,0) \qquad \{t_3\}$$

$$M_6 = M_5[t_3> = (0,0,0,1,0,0) \qquad \{\}$$

$$M_7 = M_4[t_2> = (0,0,1,0,1,0) \qquad \{\}$$

All of the possible firing sequences are

t_1
t_2,t_4,t_5,t_1,t_3
t_2,t_4,t_5,t_2

It is clearly not t_0, since any of the five transitions can fire in some sequence. It is at least L1-live since every transition can fire at least once in some sequences. It is not L2-live since only t_1 and t_2 can be fired more than once in any of the sequences. It is not L3-live since none of the transitions can be fired indefinitely. It is not L4-live since no transition, much less all transitions, can be fired indefinitely. The Petri net is nonlive since there are transitions which are nonlive. It is L1-live since all transitions are L1-live.

8.6.4 Reversibility

A Petri net is reversible if for each reachable marking M, M_0 is reachable from M

$$\forall\, M_i \in R(M_0),\; M_0 \in R(M_i)$$

In a reversible Petri net one can always get back to the initial state.

8.6.5 Fairness

Two transitions t_1 and t_2 are said to be in bounded-fair relation if the minimum number of times that either one can fire while the other is not firing is bounded. Figure 8.37 shows two Petri nets for safe traffic lights. They are both safe. Transitions rg_1 and rg_2 for the Petri net on the left are not fair. In this Petri net, rg_1 can fire an infinite number of times without allowing rg_2 to ever fire. On the other hand, transitions rg_1 and rg_2 alternate being able to fire for each cycle through the light, making the two transitions bounded-fair.

A Petri net is said to be a bounded-fair net if each pair of transitions are in a bounded-fair relation. The Petri net on the right side of Figure 8.37 bounded fair since any pair of transitions can only fire once per cycle of the traffic light.

8.6.6 Incidence Matrix

For a Petri net with n transitions and m places, the incidence matrix is n by m matrix of the form

$$A = [a_{ij}]$$

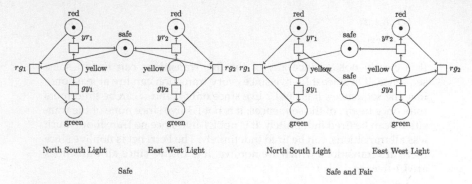

FIGURE 8.37
Fair Petri nets.

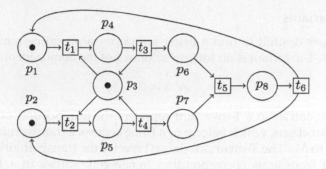

FIGURE 8.38
Example Petri net for incidence matrix.

where

$$a_{ij} = O(t_i, p_j) - I(t_i, p_j)$$

The transpose of the incidence matrix A^T is an m by n matrix in which the rows of the transpose are the columns of the incidence matrix.

Example 8.14

The first row of the incident matrix for the Petri net of Figure 8.38 is $a_{1,i} = O(t_1, p_i) - I(t_1, p_i) \rightarrow [-1\ 0\ -1\ 1\ 0\ 0\ 0\ 0]$, with the second row given by is $a_{2,i} = O(t_2, p_i) - I(t_2, p_i)$, and so on. The result is as follows:

$$A = \begin{bmatrix} -1 & 0 & -1 & 1 & 0 & 0 & 0 & 0 \\ 0 & -1 & -1 & 0 & 1 & 0 & 0 & 0 \\ 0 & 0 & 1 & -1 & 0 & 1 & 0 & 0 \\ 0 & 0 & 1 & 0 & -1 & 0 & 1 & 0 \\ 0 & 0 & 0 & 0 & 0 & -1 & -1 & 1 \\ 1 & 1 & 0 & 0 & 0 & 0 & 0 & -1 \end{bmatrix}$$

The transpose of the incident matrix A^T is the original matrix with the rows exchanged for the columns. The transpose of A is as follows:

$$A^T = \begin{bmatrix} -1 & 0 & 0 & 0 & 0 & 1 \\ 0 & -1 & 0 & 0 & 0 & 1 \\ -1 & -1 & 1 & 1 & 0 & 0 \\ 1 & 0 & -1 & 0 & 0 & 0 \\ 0 & 1 & 0 & -1 & 0 & 0 \\ 0 & 0 & 1 & 0 & -1 & 0 \\ 0 & 0 & 0 & 0 & 1 & -1 \end{bmatrix}$$

8.6.7 T-Invariants

It is sometimes desired to find a firing sequence transforming a marking M_0 back to M_0. A T-invariant is an integer solution x of the homogeneous equation

$$A^T x = 0$$

The non-zero entries in a T-invariant represent the firing counts of the corresponding transitions, which belong to a firing sequence transforming a marking M_0 back to M_0. The T-invariant doesn't specify the transition firing order.

The set of transitions corresponding to non-zero entries in a T-invariant $x > 0$ is called the support of an invariant and is denoted by $||x||$. A support is said to be *minimal* if there is no other invariant y such that $y(p) \leq x(p)$ for all p.

Given a minimal support of an invariant, there is a unique minimal invariant corresponding to the minimal support. This is called a minimal support invariant. The set of all possible minimal support invariants can serve as a generator of invariants.

Example 8.15

Using the example of the Petri net illustrated in Figure 8.38, the equation $A^T x = 0$ leads to the following simultaneous equations:

$$-x_1 + x_6 = 0$$

$$-x_2 + x_6 = 0$$

$$-x_1 - x_2 + x_3 + x_4 = 0$$

$$x_1 - x_3 = 0$$

$$x_2 - x_5 = 0$$

$$x_3 - x_5 = 0$$

$$x_5 - x_6 = 0$$

The solution to these simultaneous linear equations is $x_1 = x_2 = x_3 = x_4 = x_5 = x_6$, thus

$$A^T = \begin{bmatrix} 1 \\ 1 \\ 1 \\ 1 \\ 1 \\ 1 \\ 1 \end{bmatrix}$$

Example 8.16

The previous example only had one basic solution that could scaled by multiplying it with a constant. Some Petri nets have multiple solutions to A^Tx 0. Consider the Petri net illustrated in Figure 8.39. The transpose of the incidence matrix is

$$A^T = \begin{bmatrix} -1 & 0 & 1 & 1 \\ 1 & -1 & 0 & 0 \\ 0 & 1 & -1 & -1 \end{bmatrix}$$

The simultaneous equations are follows:

$$-x_1 + x_3 + x_4 = 0$$

$$x_1 - x_2 = 0$$

$$x_2 - x_3 - x_3 = 0$$

There are numerous solutions to these equations,

$$[1\ 1\ 0\ 1]$$

$$[1\ 1\ 1\ 0]$$

$$[2\ 2\ 1\ 1]$$

$$[3\ 3\ 2\ 1]$$

$$[3\ 3\ 1\ 2]$$

In fact, an infinity of solutions exist.

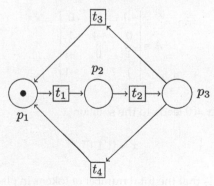

FIGURE 8.39
Second example for T-invariant.

8.6.8 S-Invariants

An S-invariant, which is also called a P-invariant, is an integer solution y of
the equation

$$Ax = 0$$

The non-zero entries in an S-invariant represent weights associated with the
corresponding places so that the weighted sum of tokens on these places is
constant for all markings reachable from an initial marking. These places are
said to be covered by an S-invariant.

Saying that a *subset* of places within a Petri net has a constant count of
tokens isn't saying that the count of tokens remains constant across the
whole Petri net, just in the particular subset. Most commonly, the goal is
to identify minimal subsets, that is, subsets that are place invariant and
yet aren't unions of other subsets that are place invariant. There may
be multiple subsets within the Petri net in which the count of tokens
remains constant. Some will be unions of other subsets that are place
invariant.

The set of places corresponding to non-zero entries in an S-invariant $x \geq 0$
is called the *support* of an invariant and is denoted by $||x||$. A support is
said to be minimal if there is no other invariant y such that $y(p) \leq x(p)$ for
all p. Given a minimal support of an invariant, this is a unique minimal
invariant corresponding to the minimal support. This is called a *minimal-
support invariant*. The set of all possible minimal support invariants can serve
as a generator of invariants!

Example 8.17

Returning to the example of Figure 8.39, the incidence matrix is as
follows:

$$A = \begin{bmatrix} -1 & 0 & 0 \\ 0 & -1 & 1 \\ 1 & 0 & -1 \\ 1 & 0 & -1 \end{bmatrix}$$

The solution of $Ax = 0$ leads to the solution

$$x = [1\ 1\ 1]^{\mathsf{T}}$$

What this means is that the total number of tokens in places p_1, p_2, and p_3
will be 1 for any marking that is reachable from M_0.

Example 8.18

The Petri net shown in Figure 8.38 has the following incidence matrix:

$$A = \begin{bmatrix} -1 & 0 & -1 & 1 & 0 & 0 & 0 & 0 \\ 0 & -1 & -1 & 0 & 1 & 0 & 0 & 0 \\ 0 & 0 & 1 & -1 & 0 & 1 & 0 & 0 \\ 0 & 0 & 1 & 0 & -1 & 0 & 1 & 0 \\ 0 & 0 & 0 & 0 & 0 & -1 & -1 & 1 \\ 1 & 1 & 0 & 0 & 0 & 0 & 0 & -1 \end{bmatrix}$$

One solution of $Ax = 0$ is $[0\,0\,1\,1\,1\,0\,0\,0]^T$. What this means is that the total number of tokens in places p_3, p_4, and p_5 in any reachable place will be the same as the total number of tokens for those places in M_0.

8.6.9 Siphons and Traps

Given a Petri net $P = (P, T, I, O, M_0)$, one can examine the behavior of what happens to a subset S of the set of states P. Let *S denote the set of input transitions into the set S, and S* denote the sets of output transitions from set S.

Example 8.19

The Petri net in Figure 8.40 can be used to demonstrate *S and S* for sets of states that are subsets of P. Let's take p_1 and p_3 as the elements of S. *S represents the input transitions into S, which are t_1 and t_4:

$$*S = \{t_1, t_4\}$$

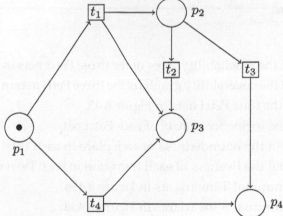

FIGURE 8.40
Sink and siphon example Petri net.

S* is the set of output transitions from S, which are t_1 (it is out from p_1), t_2 (it is output from p_1), t_3 (it is output from p_3), and p_4 (it is output from p_3):

$$S^* = \{t_1, t_2, t_3, t_4\}$$

Given a PN $N = (P, T, I, O, M_0)$, a subset of places S of P is called a *siphon* if *S $\subseteq$ S*. Intuitively, if a transition is going to deposit a token to a place in a siphon S, the transition must also remove a token from S. Think of a siphon as slowly draining the set S of tokens.

Example 8.20

For the Petri net shown in Figure 8.40, excluding the empty set the set of all possible subsets of P are

$\{p_1\}, \{p_2\}, \{p_3\}, \{p_4\}, \{p_1, p_2\}, \{p_1, p_3\}, \{p_1,p_4\}, \{p_2, p_3\}, \{p_2, p_4\}, \{p_3,p_4\}, \{p_1,p_2,p_3\},$
$\{p_1, p_2, p_4\}, \{p_1,p_3,p_4\}, \{p_2,p_3,p_4\}, \{p_1, p_2, p_3, p_4\}$

$$^*\{p_1\} = \{\}$$

$$p_1{}^* = \{t_1, t_2\}$$

Thus, $\{p_1\}$ is a siphon.

Given a Petri net $N = (P, T, I, O, M_0)$, a subset of places S of P is called a *trap* if S* $\subseteq$ *S. Intuitively, a trap S represents a set of places in which every transition consuming a token from S must also deposit a token back into S.

Exercises

1. Construct the reachability trees of the three Petri nets in Figure 8.41.
2. Construct the coverability graphs of the three Petri nets in Figure 8.42.
3. Consider the four Petri nets in Figure 8.43.
   (1) Give the incidence matrix of each Petri net.
   (2) Find out the boundedness of each place in each Petri net.
   (3) Find out the liveness of each transition in each Petri net.
4. Find the minimal T-invariants in Figure 8.43b.
5. Find the minimal S-invariants in Figure 8.43d.
6. Find the minimal T-invariants and S-invariants in the Petri net in Figure 8.44. Notice that $I(t_2, p_5) = 2$ and $O(t_3, p_5) = 3$.

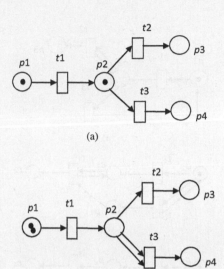

(a)

(b)

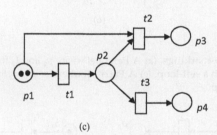

(c)

FIGURE 8.41
Three Petri nets with ω markings. (a) A Petri net with decision making, (b) a Petri net with multiple arcs from p_2 to t_3, and (c) a Petri net with synchronization.

7. In the Petri net shown in Figure 8.45, let

$$S_1 = \{p_1, p_2, p_3\}$$

$$S_2 = \{p_1, p_2, p_4\}$$

$$S_3 = \{p_1, p_2, p_3, p_4\}$$

$$S_4 = \{p_2, p_3\}$$

$$S_5 = \{p_2, p_3, p_4\}$$

Is each of the sets a siphon and/or a trap?

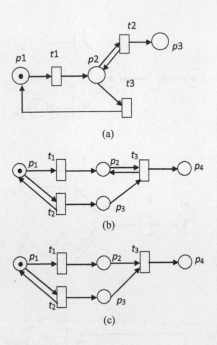

(a)

(b)

(c)

FIGURE 8.42
Three Petri nets with w-markings. (a) A Petri net where p_2 and t_2 form a self-loop. (b) A Petri net where p_1 and t_2 form a self-loop. (c) A Petri net where p_1 and t_2 form a self-loop and p_2 and t_3 form another self-loop.

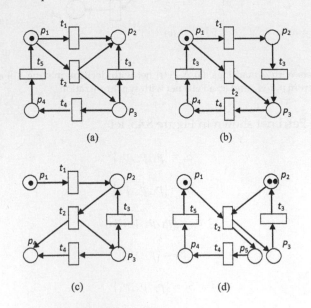

(a)

(b)

(c)

(d)

FIGURE 8.43
Four Petri nets for Problem 3. (a) The original Petri net. (b) Removed arc (t_2, p_2). (c) Removed t_5. (d) Removed t_1 and splitted p_3.

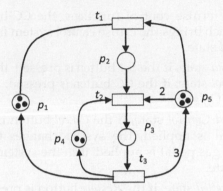

FIGURE 8.44
Petri net for Problem 6.

8. Use a Petri net to model a manufacturing system with a single machine and buffer. Events with the system include:

- A part arrives into the buffer.
- The machine starts processing.
- The machine ends processing.
- During processing, the machine may fail.
- If the machine fails, it will be repaired.
- After the machine is repaired, the processing continues.

Assume that the buffer can hold up to three parts. When the machine starts processing a part, one buffer slot is freed up for a new part.

9. Consider a cruise control (CC) system in an auto. The CC controller has four buttons:

CC, Set, Cancel, and *Resume*

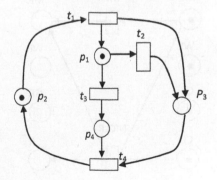

FIGURE 8.45
Petri net for Problem 7.

To start any cruise control functions, the *CC* button has to be pressed, which brings the cruise control system from the *Off* state to the *Armed* state.

- At the *Armed* state, if the *Set* button is pressed, the system enters *Speed Control* state; if the *CC* button is pressed, the system goes back to the *Off* state.

- At the *Speed Control* state, if the *Cancel* button is pressed or the brake pedal is applied, the system changes to the *Cancelled* state; if the gas pedal is applied, then the system changes to the *Override* state.

- At the *Cancelled* state, if the *Resume* button is pressed, the system goes back to the *Speed Control* state; if the *CC* button is pressed, it goes back to the *Off* state.

- At the *Override* state, if the *Resume* button is pressed, the system goes back to the *Speed Control* state; if the *CC* button is pressed, it goes back to the *Off* state; if the *Cancel* button is pressed, it switches to the *Cancelled* state.

Model the behavior of the cruise controller with a Petri net.

10. Consider the classic ferryman puzzle. A ferryman has to bring a goat, a wolf, and a cabbage from the left bank to the right bank of a river. The ferryman can cross the river either alone or with exactly one of these three passengers. At any time, either the ferryman should be on the same bank as the goat, or the goat should be alone on a bank. Otherwise, the goat will eat the cabbage or the wolf will eat the goat. In Figure 8.46, we use places ML, WL, GL, and CL to model the ferryman, wolf, goat, and cabbage on the left bank, respectively. Similarly, we use places MR, WR, GR, and CR to model the ferryman, wolf, goat, and cabbage on the right bank, respectively. Tokens

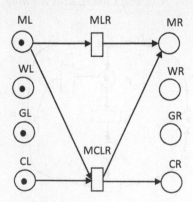

FIGURE 8.46
Ferryman crosses river (incomplete).

in MR, WR, GR, and CR indicate that initially the four agents are all on the left bank. Transition MLR models the event that the ferryman travels alone to the right bank. Transition MCLR models the event that the ferryman travels to the right bank with the cabbage.

(1) Model the event that the ferryman travels from the right bank to the left bank with the goat.

(2) Model the event that the goat eats the cabbage on the left bank. Be sure to model all preconditions and postconditions for the event.

(3) Model the event that the wolf eats the goat on the right bank. Be sure to model all preconditions and postconditions for the event.

(4) Find a sequence of transitions that enables the ferryman to bring all passengers safely to the right bank.

11. A seat belt is an important vehicle safety device that prevents the occupant of a vehicle from movement in the event of a collision or a sudden stop. A seat belt reminder system gives a signal after the ignition is turned on and if the occupant's seat belt is not fastened. Assume we want to design a seat belt reminder system that would function according to the following specifications:

- The initial state is the car engine being off.

- After being seated, the driver can either turn on the engine or put the seat belt on.

- When the engine is on but the driver is seated without seat belt buckled, the buzzer timer is turned on. The timer is turned off if the driver puts the seat belt on before the timer expires.

- If the timer expires, the buzzer is turned on. When the driver puts on his seat belt, the buzzer is turned off.

- The driver can turn off the car engine at any moment, which turns the timer or buzzer off, whichever is on.

- When the driver is seated with the belt buckled, he can take off the seat belt.

- The driver cannot turn on car engine before he is seated.

- The driver cannot leave the seat with the seat belt is buckled or while the engine is on.

Model the behavior of the seat belt system with a Petri net. Make sure your Petri net model is safe and live.

9

Timed Petri Nets

The need for including timing variables in the models of various types of dynamic systems is apparent since these systems are real-time in nature. In the real world, almost every event is time related. Traditional Petri nets, as we introduced in Chapter 8, are not able to model and analyze a system's time-related properties. Therefore, an extension of Petri nets with timing parameters is necessary. Petri nets that contain timing variables are called *timed Petri nets*. In contrast, we call Petri nets that do not have any timing parameters *regular* Petri nets. There are various types of timed Petri nets, though. This chapter introduces deterministic timed Petri nets and stochastic Petri nets.

9.1 Introducing Time to Petri Nets

A timed Petri net takes the same topological structure as its underlying Petri net model. In other words, removing all timing specifications from a timed Petri net will result in a regular Petri net. Based on the structure of Petri net models, there are three options to introducing timing parameters:

- Associate timing parameters with arcs
- Associate timing parameters with places
- Associate timing parameters with transitions

The labeling of a timed Petri net consists of assigning numerical values to transitions, places, or arcs. As far as the value of time, we also have some options. It can be deterministic, that is, we assign a fixed time to each transition, place, or arc. It can also be random, that is, we assign a time that follows certain probability distribution to each transition, place, or arc.

The fixed arc-timed, place-timed, and transition-timed Petri nets are illustrated in Figure 9.1. In an arc-timed Petri net, when a place gets a token, the token will *travel* for the time defined on the arc to the transition at the other end of the arc. In a place-timed Petri net, when a place gets a token, the token will not be available for its outgoing transitions until the time defined on the place has elapsed. In a transition-timed Petri net, when a transition is enabled, the transition will not fire until the time defined on the transition has elapsed.

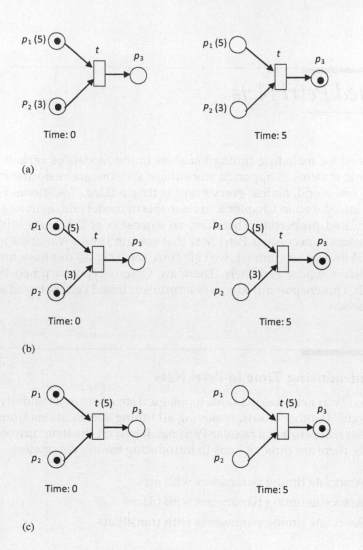

FIGURE 9.1
Introducing times to Petri nets. (a) A place-timed Petri net, (b) an arc-timed Petri net, and (c) a transition-timed Petri net.

The firing rules, which control the process of moving tokens around, are defined differently depending on the way the Petri net is labeled with timing variables and the characteristics of timing parameters. In this chapter, we only discuss timed Petri nets in which timing variables are assigned to transitions, because this is the most widely used class of timed Petri nets in both the literature and applications. Each timing parameter is a *transition firing time*, which is the duration from when a transition becomes enabled to the time it fires. It models the time that it takes for an event to complete.

9.2 Deterministic Timed Petri Nets

A *deterministic timed Petri net* (DTPN) is a 6-tuple (P, T, I, O, M_0, ST), where (P, T, I, O, M_0) is a regular Petri net, and $ST : T \to R^+$ is a function that associates each transition with a deterministic firing time, called the *static firing time* of the transition. $ST(t_i)$ is the delay of the transition t_i from being enabled to firing.

In addition to static firing time, there is also a *dynamic firing time* for each transition in DTPN. To introduce the concept of dynamic firing time, consider the simple DTPN in Figure 9.2. As a convention, we put a transition's static firing time in a pair of parentheses, right after the label of the transition. In this model, $ST(t_1) = 2$, $ST(t_2) = 7$, $ST(t_3) = 3$, and $ST(t_4) = 3$. Moreover, without loss of generality, we always assume a timed Petri net starts running at time 0 from the initial marking. The initial marking of this DTPN is

$$M_0 = (1, 1, 0, 0, 0, 0)$$

Enabled transitions are t_1 and t_2. At time 0, we know t_1 will fire at time 2 and t_2 will fire at time 7, if they are not disabled before their static firing times are up. Because $ST(t_1) < ST(t_2)$, t_1 will fire before t_2. At time 2, t_1 fires and the new marking is

$$M_1 = (0, 1, 1, 0, 0, 0)$$

in which t_2 continues to be enabled, and meanwhile t_3 also becomes enabled. Because t_3 just becomes enabled, it is due to fire 3 time units later, at time 5. For t_2, because it became enabled at time 0, thus at time 2, it has already been enabled for 2 time units; it is due to fire in 5 time units at time 7. Therefore, the situation of the two enabled transitions is as follows:

- t_3 fires in 3 time units, which is equal to its static firing time.
- t_2 fires in 5 time units, which is its *remaining* static firing time in M_1.

We call a transition's remaining static firing time in a marking the dynamic firing time, denoted by *DT*.

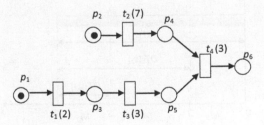

FIGURE 9.2
A simple DTPN.

9.2.1 States in DTPNs

For the transition t_i, the dynamic firing time $DT_j(t_i)$ in the marking M_j is the time from the epoch the DTPN enters M_j to the time point that t_i fires, assuming it will be enabled all the time until it can fire. Note that a transition's dynamic firing time may change from marking to marking. When a transition is newly enabled in a marking, its dynamic firing time is equal to its static firing time. If a transition was enabled in the previous marking and continued to be enabled in the current marking, then its dynamic firing time would be less than its static firing time. Figure 9.3 shows the static and dynamic firing times of t_1, t_2, and t_3 from time 0 to 7 for the DTPN shown in Figure 9.2.

In a regular Petri net, markings are the states of the system of interest, because from a marking one can decide the future behavior of the Petri net. In a DTPN model, however, markings alone are not sufficient to decide the DTPN's future behavior. Take the DTPN in Figure 9.2 as an example. In the initial marking, t_1 and t_2 are enabled. The selection of the next transition to fire depends on t_1 and t_2's static firing times, which are also their dynamic firing times; t_1 is chosen to fire because it has a shorter static firing time. Firing t_1 results in M_1, which has two enabled transitions, t_2 and t_3. Again, which one of them is selected to fire next depends on their dynamic firing times. Therefore, in a DTPN, the marking only determines if a transition can fire; all enabled transitions' dynamic firing times together decide which transition to fire next. We define a marking together with the dynamic firing times of all enabled transitions in the marking as a *state* of a DTPN. A state corresponding to M_i is denoted by

$$S_i = (M_i, DT_i)$$

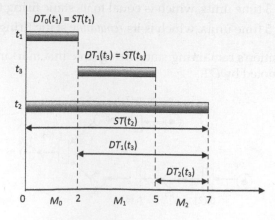

FIGURE 9.3
Static and dynamic firing times of t_1, t_2, and t_3 in Figure 9.2.

For example, the first three states of the DTPN in Figure 9.2 are

$$S_0 = (M_0, DT_0), \text{ where}$$

$$M_0: \quad (1, 1, 0, 0, 0, 0)$$

$$DT_0: \quad DT_0(t_1) = 2,$$

$$DT_0(t_2) = 7$$

$$S_1 = (M_1, DT_1), \text{ where}$$

$$M_1: \quad (0, 1, 1, 0, 0, 0),$$

$$DT_1: \quad DT_1(t_2) = 5,$$

$$DT_1(t_3) = 3$$

$$S_2 = (M_2, DT_2), \text{ where}$$

$$M_2: \quad (0, 1, 0, 0, 1, 0)$$

$$DT_0: \quad DT_2(t_2) = 2$$

9.2.2 Transition Firing Rules

Now we are ready to give a formal description of the transition firing policy of DTPNs. Consider a DTPN in the marking M_i. The transition t_k is enabled in M_i, that is,

$$M_i \geq I(t_k)$$

The next marking after a transition fires in M_i is M_j.

Preconditions:

1. t_k can fire at the time τ in the marking M_i if and only if for any input place p of t_k, there have been the number of tokens greater than or equal to $I(t_k, p)$ in p continuously for the time interval $[\tau - \tau_k, \tau]$, where $\tau_k = ST(t_k)$.

2. t_k will fire at the time τ if and only if

$$DT_i(t_k) = \min \{DT_i(t_l) | t_l \in E(M_i)\}$$

where $E(M_i)$ is the set of all enabled transitions in M_i.

Postconditions:

1. $M_j = M_i + O(t_k) - I(t_k)$

2. $DT_j(t_l) = ST_j(t_l)$, for all $t_l \in E(M_j)\backslash E(M_i)$
3. $DT_j(t_l) = DT_i(t_l) - DT_i(t_k)$, for all $t_l \in E(M_i) \cap E(M_j)$

The first postcondition is the same as it is with regular Petri nets. The second precondition states that for all newly enabled transitions in the new marking, their dynamic firing times are equal to their static firing times. The third postcondition states that for a transition that was enabled before the firing and continues to be enabled in the new marking, its dynamic firing time in the new state is the one in the old state minus the firing transition's dynamic firing time in the previous state.

A transition will fire and has to fire when the precondition is satisfied. This erases some of the uncertainty that regular Petri nets behave. Recall in the case that a marking enables multiple transitions in a regular Petri net, which enabled transition should fire first is undecided. Therefore, when we produce the reachability tree, we consider all possibilities. This is not the case with DTPN. The transition that reaches its firing time sooner than any other enabled transition is the next transition to fire. That is why we fire t_1 first in the previous example, even with t_2 being enabled as well. In particular, in a decision case like the one shown in Figure 9.4 where three transitions compete for a single token from a place, if they all have different firing times, it will be always the one with the shortest firing time to fire, while the other two transitions will never get fired. In general, the state space of a DTPN is only a subset of its underlying regular Petri net.

Example 9.1: A DTPN with Resource Sharing

We use the DTPN in Figure 9.5 as an example to show the step-by-step state changes of a DTPN. Transitions' static firing times are shown in the model. The initial marking is $(1, 0, 0, 0, 0, 0, 0, 0, 0, 0, 1)$. Because this is a safe Petri net, we will denote a marking by the names of places that contain a token to improve readability. As such, we denote M_0 by $p_1 p_{11}$.

The entire state changes of this model are illustrated in Figure 9.6. The transition firing sequence is

$$t_1\ t_2\ t_4\ t_3\ t_5\ t_6\ t_7\ t_8$$

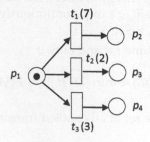

FIGURE 9.4
A decision case.

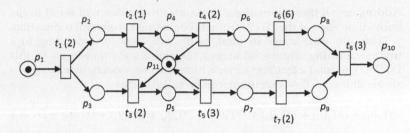

FIGURE 9.5
A DTPN with resource sharing.

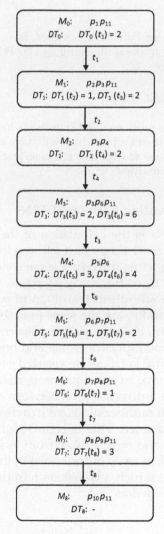

FIGURE 9.6
State transitions of the DTPN in Figure 9.5.

Adding up all these transitions' dynamic firing times will result in the entire time span of the DTPN from its beginning to run until completion.

There are nine states in total, with each state corresponding to a unique marking. Figure 9.6 shows the entire state transitions of this DTPN. The total execution time, which often represents time span of a single shot workflow execution, can be calculated as

$$DT_0(t_1) + DT_1(t_2) + DT_2(t_4) + DT_3(t_3) + DT_4(t_5) + DT_5(t_6) + DT_6(t_7) + DT_7(t_8)$$

$$= 2 + 1 + 2 + 2 + 3 + 1 + 1 + 3$$

$$= 15$$

The underlying regular Petri net, however, has more markings. The markings that appear in the underlying regular Petri net model but not in the DTPN model are

$$p_2 p_5; \quad p_2 p_7 p_{11}; \quad p_4 p_7; \quad p_6 p_9 p_{11}$$

If in a state the two or more enabled transitions have the same smallest dynamic firing time, then each of them will have an equal opportunity to fire, and thus we have to explore each case, just as we did in regular Petri net reachability analysis.

9.2.3 Performance Evaluation Based on DTPNs

One application of DTPNs is the calculation of the cycle time of a class of systems in which job arrival times and job service times are known and fixed. Before we proceed, let us introduce some concepts first. In a Petri net, a sequence of places and transitions, say $p_1 t_1 p_2 t_2 \ldots p_{k-1} t_{k-1} p_k$, is a *directed path* from p_1 to p_k if t_i is both an output transition of p_i and input transition of p_{i+1} for $1 \le i \le k - 1$. If p_1 and p_k are the same place but all other nodes in the directed path are different, then the path is a *directed circuit*. If in a Petri net every place has exactly one input transition and one output transition, then the Petri net is a *decision-free* Petri net or a *marked graph*.

Decision-free DTPNs have two unique properties. First, they are *strongly connected*, that is, there is a directed path between any two nodes in such a Petri net. Second, the total number of tokens in a directed circuit remains the same after any firing sequence. This is because whenever a transition in a circuit fires, it removes 1 and only 1 token from its input place in the circuit and adds 1 and only 1 token to its output place in the circuit.

Let $S_i(n_i)$ be the time at which a transition t_i initiates its n_i-th firing. Then the *cycle time* C_i of t_i is defined as

$$C_i = \lim_{n_i \to \infty} \frac{S_i(n_i)}{n_i}$$

It is proved that *all* transitions in a decision-free DTPN have the same cycle time C. Here we explain how to calculate C. Consider a decision-free DTPN with q directed circuits. For a circuit L_k, denote the sum of the firing times of all transitions in the circuit by T_k and the total number of tokens in all places of the circuit by N_k, that is,

$$T_k = \sum_{t_i \in L_k} ST(t_i)$$

$$N_k = \sum_{p_i \in L_k} M(p_i)$$

Both T_k and N_k are constants. N_k can be counted at the initial marking. Obviously, the number of transitions that are enabled simultaneously in L_k is less than or equal to N_k. On the other hand, the processing time required by circuit L_k per cycle, which is T_k, is less than or equal to the maximum processing power of the circuit per cycle time, which is denoted by CN_k. Therefore, we have

$$T_k \leq CN_k$$

or

$$C \geq \frac{T_k}{N_k}$$

The bottleneck circuit in the decision-free Petri net is the one that satisfies

$$T_k = CN_k$$

Therefore, the minimum cycle time C is given by

$$C = \max\{\frac{T_k}{N_k} \mid k = 1, 2, \ldots, q\}$$

which corresponds to the best performance of the system modeled by the DTPN.

Example 9.2: DTPN-Based Performance Evaluation

Consider the DTPN shown in Figure 9.7. Its underlying regular Petri net is a decision-free net. We want to find the minimum cycle time of this DTPN. There are four circuits:

L_1: $p_4 t_4 p_8 t_2 p_4$
L_2: $p_5 t_5 p_9 t_3 p_5$
L_3: $p_1 t_1 p_2 t_2 p_4 t_4 p_6 t_6 p_{10} t_7 p_{11} t_8 p_1$
L_4: $p_1 t_1 p_3 t_3 p_5 t_5 p_7 t_6 p_{10} t_7 p_{11} t_8 p_1$

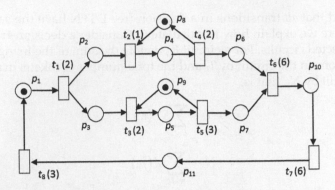

FIGURE 9.7
A decision-free DTPN.

The calculation of the cycle time of each circuit is as follows:

$$C_{L_1} = \frac{T_1}{N_1} = \frac{2+1}{1} = 3$$

$$C_{L_2} = \frac{T_2}{N_2} = \frac{3+2}{1} = 5$$

$$C_{L_3} = \frac{T_3}{N_3} = \frac{2+1+2+6+6+3}{1} = 20$$

$$C_{L_4} = \frac{T_4}{N_4} = \frac{2+2+3+6+6+3}{1} = 22$$

The minimum cycle time of the DTPN is

$$C = \max\{3,5,20,22\} = 22$$

9.3 Probability and Stochastic Process

In this section, we review the basic concepts of probability and stochastic process. An important topic is Markov process, which is the mathematic basis of stochastic Petri nets that will be presented in Section 9.4.

9.3.1 Probability

Probability is the measure of the likelihood that an event will occur. Envision an experiment for which the result is unknown. The collection of all possible *outcomes* is called the *sample space* of the experiment. An outcome is also called a *sample point*. A set of outcomes, or subset of the sample space, is called an *event*.

The core concept in probability theory is that of a *probability model*. A probability model is a 3-tuple $(\Omega, \mathcal{E}, \mathrm{Pr})$ where Ω is a sample space, $\mathcal{E}$ is a collection of events from the sample space, and Pr is a probability law that assigns a real number to each event in $\mathcal{E}$.

For example, rolling a dice can produce six possible results, that is,

$$\Omega = \{1, 2, 3, 4, 5, 6\}$$

If we are only interested in whether the result on the dice is an odd number or even number, then there are two events here. The *odd* event is the subset $\{1, 3, 5\}$, while the *even* event the subset $\{2, 4, 6\}$. That is,

$$\mathcal{E} = \{\text{Odd, Even}\}$$

If the outcomes that actually occur fall in a given event, the event is said to have occurred. If the dice has an equal chance to be on each of the numbers when it is rolled, then the probability for each outcome is 1/6 and that for each event is 1/2. That is,

$$\mathrm{Pr}\{\text{outcome} = \text{odd number}\} = \mathrm{Pr}\{\text{outcome} = \text{even number}\} = 0.5$$

A random variable is a real function defined over a probability space. For a given number x, there is a fixed possibility that a random variable X will not exceed this value, written as $\mathrm{Pr}\{X \leq x\}$. The probability is a function of x, known as $F_X(x)$.

$$F_X(x) = \mathrm{Pr}\{X \leq x\}$$

$F_X(x)$ is the *cumulative distribution function* (CDF) of X. It is nonnegative and nondecreasing over x, and satisfying

$$\lim_{x \to -\infty} F_X(x) = 0$$

$$\lim_{x \to +\infty} F_X(x) = 1$$

A *continuous* random variable has a *probability density function* (pdf), which is defined as

$$f_X(x) = \frac{dF_X(x)}{dx}$$

A *discrete* random variable takes values in a discrete set. It does not have PDF, but we consider its *probability mass function*, which is defined as

$$p_X = (p_1, p_2, \ldots)$$

where

$$p_i = \Pr\{X = x_i, i = 1, 2, \ldots\}$$

The joint CDF of random variables X_1, X_2, ..., X_n is given by

$$F_X(x) = \Pr\{X_1 \le x_1, X_2 \le x_2, \ldots, X_n \le x_n\}$$

If X_1, X_2,..., X_n are independent, then

$$F_X(x) = \Pr\{X_1 \le x_1\} \cdot \Pr\{X_2 \le x_2\} \cdot \ldots \cdot \Pr\{X_n \le x_n\}$$

If A and B are events in $\mathcal{E}$ with $\Pr\{B\} \ne 0$, the *conditional probability* of A given that B has occurred is

$$\Pr\{A \mid B\} = \frac{\Pr\{A, B\}}{\Pr\{B\}}$$

Example 9.3: Exponential Distribution

The *exponential distribution*, also known as the *negative exponential distribution*, is the probability distribution that can specify the time between customers who visit a barber shop, the time between iPhone users who make purchases on the App Store, and so on. These kinds of events (visits and purchases) occur continuously and independently at a constant average rate. Assume a random variable X has an exponential distribution and its average rate is denoted by λ. Then the pdf of X is

$$f_X(t) = \begin{cases} \lambda e^{-\lambda t} & t \ge 0, \\ 0 & t < 0 \end{cases}$$

The CDF is given by

$$F_X(t) = \begin{cases} 1 - e^{-\lambda t} & t \ge 0, \\ 0 & t < 0 \end{cases}$$

The pdf and CDF are shown in Figure 9.8.

A very useful property of an exponential distribution is that it is *memoryless*, that is

$$\Pr\{X > s + t \mid X > s\} = \Pr\{X > t\}, \forall s, t \ge 0$$

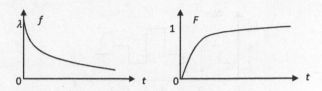

FIGURE 9.8
Illustration of pdf and CDF of exponential distribution.

If X is interpreted as the waiting time for a light bulb to fail and it has an exponential distribution, then the memoryless property states that the distribution of remaining waiting time that a failure of the bulb to occur conditioned on that the failure didn't occur in (0, s), is the same as the original unconditional distribution. In other words, if you have a light bulb that you have used for a period of time and it still works, then its remaining life span is the same as that when it was brand new.

This property can be easily proved from the definition of the distribution. The proof is left as an exercise.

9.3.2 Stochastic Process

A *stochastic process* is a collection of random variables indexed on some mathematical set. The index is often interpreted as time. In that case, each sample point or outcome maps to a function of time is called a *sample path*. A simple path is a single outcome of a stochastic process and formed by taking a single possible value of each random variable. A sample path might vary continuously with time or might vary only at discrete times. If it varies at discrete times, those times can be either deterministic or random. Besides, each random variable in the collection takes values from the same mathematical space known as the *state space*. This state space can be, for example, integers.

A continuous-time stochastic process is denoted by

$$\{X_t, t \geq 0\}$$

in which X_t can be continuous-valued or discrete-valued. A discrete time stochastic process is denoted by

$$\{X_n, n = 0, 1, 2, \ldots\}$$

For example, a *counting process* is a discrete-valued, continuous-time stochastic process that increases by one each time some event occurs. The value of the process at time t is the number of events that have occurred up to (and including) time t.

The sample points of the probability model can be taken to be the sample paths of the process.

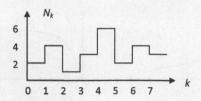

FIGURE 9.9
A sample path of the process of rolling a dice repeatedly.

Example 9.4: Rolling a Dice Repeatedly

Consider an experiment that at time instants $k = 0, 1, 2, \ldots$, we roll a dice and record the outcome N_k, $N_k \in \{1, 2, 3, 4, 5, 6\}$. This constitutes a stochastic process $\{X_k, k = 0, 1, 2, \ldots\}$, where $X_k = N_k$. The set $\{1, 2, 3, 4, 5, 6\}$ forms the state space of this process. A sample path is shown in Figure 9.9.

9.3.3 Continuous-Time Markov Chains

Consider a continuous-time stochastic process $\{X(t), t \geq 0\}$ that takes values from the set of nonnegative integers, that is, the process's state space is

$$S = \{0, 1, 2, \ldots\}$$

If the process satisfies the following *Markovian property*

$$\Pr\{X(t+s) = j \mid X(s) = i, X(t_{n-1}) = i_{n-1}, \ldots, X(t_1) = i_1\}$$

$$= \Pr\{X(t+s) = j \mid X(s) = i\}, \quad \forall t \geq \ s \geq t_{n-1} \geq \ldots \geq t_1 \geq t_0$$

we call it a *continuous-time Markov chain* (CTMC). The Markovian property indicates that a Markov process's behavior in the future (at some time $t + s$) depends on only the present state (at time s), but not on the history (at $t_{n-1}, \ldots$ t_1, t_0). Thus, this property is also called *memoryless property*.

If, in addition, the probability

$$\Pr\{X(t+s) = j \mid X(s) = i\}, \quad \forall t \geq s$$

is independent of s, then the CTMC is said to have *stationary* or *homogeneous* transition probabilities.

Suppose that a CTMC enters state i at some time, say time 0, and the process does not leave state i (i.e., a transition does not occur) during the next s time units. What is the probability that the process will not leave state i during the following t time units? To answer this question, note that as the process is in state i at time s, it follows, by Markovian property, that the probability it remains in that state during the interval $[s, s+t]$ is just the (unconditional) probability that it stays in state i for at least t time units. That is, if we let τ_i denote the amount of time that the process stays in state i before making a transition into a different state, then

$$\Pr\{\tau_i > s + t \mid \tau_i > s\} = \Pr\{\tau_i > t\}$$

for all $s, t \geq 0$. Hence, the random variable τ_i is memoryless and must be exponentially distributed. This fact tells us how to construct a CTMC. Namely, a CTMC is a stochastic process having the properties that each time it enters state i:

- The amount of time it spends in that state before making a transition into a different state is exponentially distributed.
- When the process leaves state i, it will enter state j with some probability, say p_{ij}, such that

$$\sum_{j \neq i} p_{ij} = 1$$

A CTMC is said to be *irreducible* if it can go from any state to any other state. A state i is said to be *transient* if, given that the CTMC starts in i, there is a non-zero probability that it will never return to i. State i is *recurrent* (or persistent) if it is not transient. It is proved that for an irreducible and positive recurrent homogeneous CTMC, the limiting probabilities

$$\pi_i = \lim_{t \to \infty} p_i(t)$$

always exist, where

$$p_i(t) = \Pr\{X(t) = i\}$$

Moreover, these limiting probabilities constitute a probability distribution, that is,

$$\sum_{i \in S} \pi_i = 1$$

The probabilities π_i, $i \in S$, can be interpreted as the long-run proportion of residence time in state i. They are unique, that is, independent of the initial state. The limiting probabilities are stationary in the sense that if the initial state is chosen as

$$\Pi(0) = \Pi = (\pi_0, \pi_1, \ldots)$$

then the probabilities $\Pi(t)$ at any time $t > 0$ will be the same as $\Pi(0)$. Therefore, the probabilities are said to constitute the *steady-state probability distribution* of the CTMC.

A CTMC is characterized by either a *state transition rate diagram* or a *transition rate matrix*; the latter is also called *infinitesimal generator matrix* and denoted by Q. The state transition rate diagram is a labeled directed graph whose vertices are labeled with the CTMC states and whose arcs are labeled with the rate of the exponential distribution associated with the transition from a state to another. The infinitesimal generator is a matrix whose elements outside the main diagonal are the rates of the exponential distributions associated with the transitions from a state, while the elements on the main diagonal make the sum of the elements of each row equal to zero.

With the infinitesimal generator Q, the steady-state probability vector Π can be obtained as the solution of

$$\Pi Q = 0$$

$$\sum_{i \in S} \pi_i = 1$$

If the state space S is finite, say $S = \{0, 1, ..., n - 1\}$, then Π is an $1 \times n$ vector, while Q is an $n \times n$ matrix. An individual term of $\Pi Q = 0$ is given by

$$q_{ii}\pi_i + \sum_{j \in S,\, j \neq i} q_{ij}\pi_i = 0$$

where

$$\sum_{j \in S} q_{ij} = 0$$

Example 9.5: A Continuous-Time Markov Chain

Consider a manufacturing machine. When it is free, the machine may load and process a raw part. After the completion of the processing, the machine is free again. When the machine is processing a part, it may fail and thus needs repairing. After the completion of the repairing, the machine is free and waits for the arrival of a new raw part.

We can easily identify three states in this system: (1) idle, (2) processing, and (3) repairing. The transitions from state to state obey the following rules:

- When the machine is idle, it may load and process a part.
- When the machine is processing a part, it may complete the activity, or fail.
- When the machine fails, it goes for repair. When repair is done, it becomes idle.

In order to obtain a CTMC model we need to introduce temporal specifications such that the evolution in the future depends only on the present state, not on the history. To the end, we make the following assumptions:

- The time period during which the machine is idle is exponentially distributed with a rate β.
- The time period during which the machine is processing a raw material is exponentially distributed with a rate α.
- The lifetime of the machine is exponentially distributed with parameter μ.
- The repair time is exponentially distributed with a rate λ.

The state transition diagram of the resulting CTMC is depicted in Figure 9.10, in which states 0, 1, and 2 represent the idle, processing, and repairing states, respectively. The infinitesimal generator matrix is

$$Q = \begin{bmatrix} -\alpha & \alpha & 0 \\ \beta & -(\beta + \lambda) & \lambda \\ \mu & 0 & -\mu \end{bmatrix}$$

It is an irreducible and positive recurrent homogeneous CTMC, and the steady-state probability distribution is computed by solving the following system of linear equations:

$$-\alpha \pi_0 + \beta \pi_1 + \mu \pi_2 = 0$$

$$\alpha \pi_0 - (\beta + \lambda) \pi_1 = 0$$

$$\lambda \pi_1 - \mu \pi_2 = 0$$

$$\pi_0 + \pi_1 + \pi_2 = 1$$

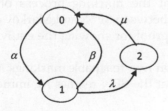

FIGURE 9.10
State transition diagram of the CTMC in Example 9.5.

The result is

$$\pi_0 = \frac{1}{\mu(\alpha+\beta)+\lambda(\mu+\alpha)}\mu(\beta+\lambda)$$

$$\pi_1 = \frac{1}{\mu(\alpha+\beta)+\lambda(\mu+\alpha)}\mu\alpha$$

$$\pi_2 = \frac{1}{\mu(\alpha+\beta)+\lambda(\mu+\alpha)}\lambda\alpha$$

Let $\lambda = \mu = 1$ and $\alpha = \beta = 2$. Then we have

$$\Pi = \left(\frac{3}{7}, \frac{2}{7}, \frac{2}{7}\right)$$

9.4 Stochastic Petri Nets

The simplest class of stochastic timed Petri nets is *stochastic Petri nets* (SPN), in which every transition is associated with exponentially distributed firing time. Because of that, the marking process of an SPN is a CTMC, which makes SPN model-based system performance evaluation relatively easy.

9.4.1 Definition

An SPN is a 6-tuple $(P, T, I, O, M_0, \Lambda)$, in which

- (P, T, I, O, M_0) is a regular Petri net.
- $\Lambda: T \to R$ is a set of firing rates whose entry λ_k is the rate of the exponential individual firing time distribution associated with transition t_k.

λ_k may be marking dependent, that is, λ_k may take different values in different markings.

Now let us show that the marking process of an SPN is a CTMC. This establishes a bridge between SPNs and Markov chains. In the following discussion, we outline a proof for showing the equivalence between an SPN and a CTMC.

First we show that given two reachable markings M_i and M_j, there exists a specific probability α_{ij} that the SPN reaches M_j immediately after leaves M_i. Let

$$T_{ij} = \{t \in E(M_i) \mid M_i[t > M_j\}$$

which is the set of all enabled transitions such that when any one of them fires the SPN moves from M_i to M_j. If $T_{ij} = \varnothing$, meaning there is no single step transition from M_i to M_j, then $\alpha_{ij} = 0$. Consider the case when $T_{ij} \neq \varnothing$. Let

$$r_{ij} = \sum_{t_k \in T_{ij}} \lambda_k$$

$$r_i = \sum_{t_k \in E(M_i)} \lambda_k$$

where r_{ij} is the total rate of the enabled transitions that will take the SPN to M_j, and r_i is the total rate of all enabled transitions in M_i that will take the SPN to any possible markings. Since the firing times in an SPN are mutually independent exponential random variables, the probability of the SPN moves to M_j in the next step is

$$\alpha_{ij} = \frac{r_{ij}}{r_i}$$

Now we show that the sojourn time of any reachable marking in an SPN is exponentially distributed. Let $E'(M_i)$ be the subset of $E(M_i)$ comprising all transitions such that firing any one in $E'(M_i)$ would take the SPN to a marking other than M_i. Let

$$r_i' = \sum_{t_k \in E'(M_i)} \lambda_k$$

Because all transitions have independent firing times, the sojourn time, or the time when the SPN leaves M_i, is also exponentially distributed and the rate is r_i'.

The above analysis not only proves that the marking process of an SPN is a CTMC, but also shows how to calculate the Q matrix of the CTMC from given transition firing rates.

Example 9.6: Steady-State Analysis of an SPN Model

Figure 9.11 shows a simple SPN model and its reachable graph. The markings in the graph are listed as follows:

M_0: (1, 0, 0, 0, 0)
M_1: (0, 1, 1, 0, 0)
M_2: (0, 0, 1, 1, 0)
M_3: (0, 1, 0, 0, 1)
M_4: (0, 0, 0, 1, 1)

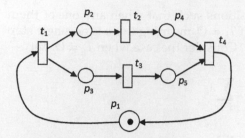
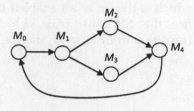

FIGURE 9.11
A simple SPN model and its reachable graph.

The linear system of $\Pi Q = 0$ is given as

$$
(\pi_0, \pi_1, \pi_2, \pi_3, \pi_4)
\begin{bmatrix}
-\lambda_1 & \lambda_1 & 0 & 0 & 0 \\
0 & -(\lambda_2 + \lambda_3) & \lambda_2 & \lambda_3 & 0 \\
0 & 0 & -\lambda_3 & 0 & \lambda_3 \\
0 & 0 & 0 & -\lambda_2 & \lambda_2 \\
\lambda_4 & 0 & 0 & 0 & -\lambda_4
\end{bmatrix} = 0
$$

which results in

$$-\pi_0 \lambda_1 + \pi_4 \lambda_4 = 0$$

$$\pi_0 \lambda_1 - \pi_1 (\lambda_2 + \lambda_3) = 0$$

$$\pi_1 \lambda_2 - \pi_2 \lambda_3 = 0$$

$$\pi_1 \lambda_3 - \pi_3 \lambda_2 = 0$$

$$\pi_2 \lambda_3 + \pi_3 \lambda_2 - \pi_4 \lambda_4 = 0$$

Also consider

$$\pi_0 + \pi_1 + \pi_2 + \pi_3 + \pi_4 = 1$$

Assume $\Lambda = (1\ 1\ 1\ 1\ 1)$. Then solution to this system is as follows:

$$\pi_0 = \pi_4 = \frac{2}{7}, \quad \pi_1 = \pi_2 = \pi_3 = \frac{1}{7}$$

Example 9.7: SPN Modeling

The SPN model describing the behavior of the manufacturing machine specified in Example 9.5 is shown in Figure 9.12. Table 9.1 lists the legend of places and transitions and transitions' firing rates in this model. By reachability analysis one can easily obtain the state (marking) transition rate diagram of this model, which is exactly same as the one shown in Figure 9.10.

9.4.2 Performance Evaluation

With the SPN model of a system of interest in place, we can analyze some performance indices of the system, such as the utilization of machines or servers, average queuing time in a system, the mean throughput of a system, and so on. We use Example 9.8 to show how this can be done.

Example 9.8: A Client-Server System

Consider a client-server system that has two servers and three client terminals. After an exponentially distributed time delay, any terminal may submit a request to the two servers for service. The two servers

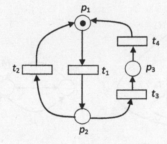

FIGURE 9.12
SPN model of the system in Example 9.5.

TABLE 9.1

Legend for Figure 9.12

Place	Description	
p_1	The machine is available.	
p_2	The machine is processing a raw part.	
P_3	The machine is under repair.	
Transition	**Description**	**Firing Rate**
t_1	The machine loads a raw part.	α
t_2	The machine processes a raw part.	β
t_3	The machine fails.	λ
t_4	The machine is repaired.	μ

have the same functions but different processing speeds. However, both service times are exponential. The SPN model of this system is shown in Figure 9.13. The legends of places and transitions and the firing rate of each transition in this model are given in Table 9.2. Notice that the firing rate of transition t_1 is marking dependent. Each single client issues service request at the rate of λ. When there are n tokens in p_1, meaning n clients are ready to issue service requests, the actual firing rate of t_1 will be $n\lambda$. We use $m(p_1)$ to represent the number of tokens in p_1 in a marking.

There are four reachable markings in this SPN:

$$M_0: (3, 0) \quad M_1: (2, 1) \quad M_2: (1, 2) \quad M_3: (0, 3)$$

The marking process constitutes a CTMC, which is also shown in Figure 9.13. Notice that the transition rate from M_0 to M_1 is 3λ because there are 3 tokens in p_1 in M_0. The transition rates from M_1 to M_2 and from M_2 to M_3 are decided the same way.

Let

$$\rho = \frac{\lambda}{\mu_1 + \mu_2}$$

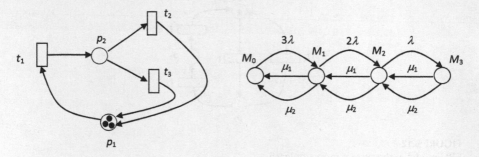

FIGURE 9.13
The SPN of a client-server system and its corresponding state transition rate diagram.

TABLE 9.2

Legend for Figure 9.13

Place	Description	
p_1	Client terminals ready to submit requests	
p_2	Requests waiting for processing	

Transition	Description	Firing Rate
t_1	Client terminals submit requests	$m(p_1)*\lambda$
t_2	Server I processes requests	μ_1
t_3	Server II processes requests	μ_2

Then steady-state probability distribution of the system is as follows:

$$\pi_0 = \frac{1}{1+3\rho+6\rho^2+6\rho^3}$$

$$\pi_1 = \frac{3\rho}{1+3\rho+6\rho^2+6\rho^3}$$

$$\pi_2 = \frac{6\rho^2}{1+3\rho+6\rho^2+6\rho^3}$$

$$\pi_3 = \frac{6\rho^3}{1+3\rho+6\rho^2+6\rho^3}$$

Now we consider the performance indices of the system.

Utilization. Since the system is idle only in the state M_0 (p_1 contains 3 tokens, indicating that no client terminal is under service), the stationary probability of the servers being idle is equal to π_0. In other words, the utilization of the two servers is

$$U = 1 - \pi_0 = \frac{3\rho+6\rho^2+6\rho^3}{1+3\rho+6\rho^2+6\rho^3}$$

Average throughput. The average throughput of the system is the average service rate, which is $\mu_1 + \mu_2$, times the server utilization. That is

$$Th = U(\mu_1 + \mu_2)$$

Average turnaround time. The average turnaround time is defined the time needed for one cycle of all three clients' requests to be served. It is calculated as

$$Tu = \frac{3}{Th}$$

Average waiting time. The average waiting time is the average turnaround time minus the average service request delay and average service time. That is,

$$W = Tu - \frac{1}{\lambda} - \frac{1}{\mu_1 + \mu_2}$$

Assume $\lambda = 2$, $\mu_1 = 1$, and $\mu_2 = 2$. Then we have $U = 0.836$, $Th = 2.509$, $Tu = 1.196$, and $W = 0.362$.

Exercises

1. Find all states of the DTPN in Figure 9.14 and draw the state transition graph.

2. Find all states of the DTPN in Figure 9.15 and draw the state transition graph.

3. Consider the decision-free DTPN in Figure 9.16.

   a.   List all directed circuits.

   b.   Calculate the minimum cycle time. Which circuit is the bottleneck circuit in terms of the processing rate?

   c.   Add a token to p_4. Recalculate the minimum cycle time.

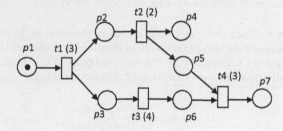

FIGURE 9.14
The DTPN for Problem 1.

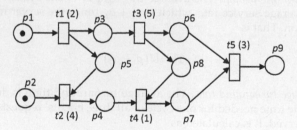

FIGURE 9.15
The DTPN for Problem 2.

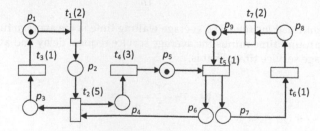

FIGURE 9.16
The DTPN for Problem 3.

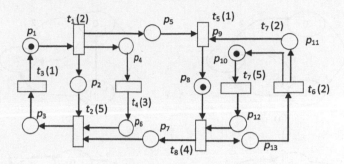

FIGURE 9.17
The DTPN for Problem 4.

4. Consider the decision-free DTPN in Figure 9.17.
   a. List all directed circuits.
   b. Calculate the minimum cycle time.

5. An infinitesimal generator matrix defines a CTMC. Find the steady-state probability distribution of each CTMC that has the following infinitesimal generator matrix.

   a. $Q = \begin{bmatrix} -1 & 1 & 0 \\ 2 & -3 & 1 \\ 1 & 0 & -1 \end{bmatrix}$

   b. $Q = \begin{bmatrix} -2 & 0 & 2 \\ 0 & -1 & 1 \\ 3 & 2 & -5 \end{bmatrix}$

   c. $Q = \begin{bmatrix} -2 & 0 & -2 \\ 1 & -2 & 1 \\ 3 & 2 & -5 \end{bmatrix}$

6. A CTMC can be specified with a state transition rate diagram in which each node is a state and each arc shows state transition and rate. Find the steady-state probability distribution of each CTMC whose state transition graph is depicted in Figure 9.18.

7. Figure 9.19 shows an SPN and its transition firing rates. Draw its state transition rate diagram and find out its steady-state probability distribution.

8. Figure 9.20 shows an SPN and its transition firing rates.
   a. Draw its state transition rate diagram and find out its steady-state probability distribution.
   b. Add one more token to p_1 and perform the steady-state probability analysis again.

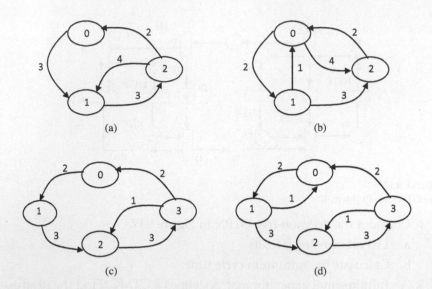

FIGURE 9.18
CTMC state transition rate graphs. (a) A three-state CTMC, (b) a three-state CTMC with more state transitions, (c) a four-state CTMC, and (d) a four-state CTMC with more state transitions.

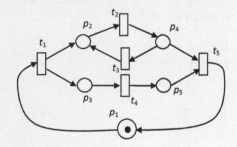

FIGURE 9.19
An SPN with $\lambda_1 = \lambda_5 = 2$ and $\lambda_2 = \lambda_3 = \lambda_4 = 1$.

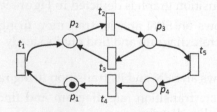

FIGURE 9.20
An SPN with $\lambda_1 = \lambda_4 = \lambda_5 = 1$ and $\lambda_2 = \lambda_3 = 2$.

10

Colored Petri Nets

The Petri nets that we introduced in Chapter 8 are often called place/transition nets, or *P/T-nets* for short. In a P/T-net, a token in a place may indicate that the condition associated with that place is satisfied. Tokens may also be seen as objects in a pool or resources in storage. When a place contains multiple tokens, however, these tokens are not distinguishable from each other. This often leads to a large size of model, because if you want to specify multiple objects differently, even if they are of the same type, you have to model each object as a token in a different place.

A colored Petri net (CPN) has each token attached with a color, indicating the identity of the token (object). Moreover, each place is attached a set of colors and each transition is associated with a set of bindings that specify the colors of tokens to take from and put into places. A transition can fire with respect to each of its bindings. By firing a transition, tokens are removed from the input places and added to the output places in the same way as that in original P/T-nets, except that a functional dependency is specified between the color of the transition firing and the colors of the involved tokens. The color attached to a token may be changed by a transition firing, and it often represents a complex data value. CPNs lead to compact net models due to the use of colored tokens.

10.1 Introductory Examples

In this section, we use three modeling examples to informally introduce CPNs. Important concepts, such as color set, binding, and transition firing guard, will be explained in these examples.

Example 10.1: A Three-Bit Binary Counter

Consider a three-bit binary counter. It has eight possible output values:

000 001 010 011 100 101 110 111

A P/T-net that models the running of the counter would need eight places, each representing a unique output value. The model is shown in Figure 10.1. It looks silly. Imagine if you want to model the behavior of

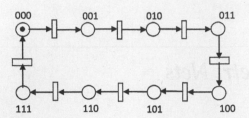

FIGURE 10.1
P/T-net model of a three-bit binary counter.

an eight-bit counter or 16-bit counter, then you would need 256 places or 65,536 places in your model, respectively.

Now consider a place that contains a *colored* token, that is, the token carries values. Define the following color (value) set:

$$OUTPUT = \{000, 001, 010, 011, 100, 101, 110, 111\}$$

and the token takes one value from the set at any given marking of the net. The place is illustrated in Figure 10.2, in which "counter" is the place-name, "OUTPUT" is the color set of the token in the place, and the "000" inside the place denotes that the place currently has a token and the token color is 000. With this place, we merge all eight places in Figure 10.1, and it results in the new model shown in Figure 10.3.

The model in Figure 10.3 has one place and eight transitions. Each arc in this model has a token color associated with it, which indicates the color of tokens flowing through the arc. For example, the input arc of the transition t_1 from the place counter is associated with 000, and the

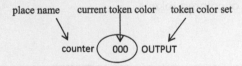

FIGURE 10.2
A place with a colored token.

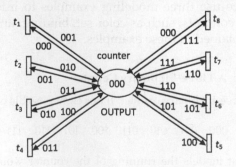

FIGURE 10.3
Model of the three-bit binary counter with a colored token.

output arc of t_1 to counter is associated with 001, which means when t_1 fires, it removes a token of color 000 from counter and adds a token of color 001 to counter. Therefore, after the firing of t_1, the original token of color 000 in counter is replaced by a token of color 001. Notice that t_1 is the only enabled transition in the initial marking of the model, because any other transition would require an input token of color other than 000 from counter. After firing t_1, the next enabled transition will be t_2, and firing t_2 will trigger t_3, and so on. It is easy to conclude that this model is equivalent to the model in Figure 10.1.

Although we merged eight places into one, the model in Figure 10.3 is still less pretty, because it has eight identical transitions. Can we use one transition to replace these eight transitions? The answer is yes. We just need to use a *variable* to specify the color of the token that the transition takes from the place, and also use another variable to indicate the color of the token that is added to the place when the transition fires. This model is shown in Figure 10.4. Here, the variable x is associated with the transition's input arc and y associated with its output arc. We say x and y *bind* to the color of the token in counter. In the initial marking, the token color is 000, and thus $x = 000$. When the transition files, it removes the token 000, and adds a token of color y to counter. Of course, we don't want a token of random color to be added to counter. Notice that in the model a formula

$$y = (x + 1) \% 8$$

where % is the modulus operator; it is associated with the transition and it specifies the relation between x and y. The formula is a *guard* of the transition, which regulates the input and output of the transition firing. The variable x always takes the color of the token in counter at the time being. When the transition files, the token is replaced by a new token whose color is the previous token color incremented by 1, which wraps back to 000 if the previous token value is 111.

Comparing the model in Figure 10.4 to its equivalent in Figure 10.1, we see a dramatic change is size. This is exactly the beauty of CPNs.

Example 10.2: A Manufacturing System

Now let us turn to another example that is a little bit "larger." Consider a manufacturing system comprising two machines that process three different types of raw parts. Each part type goes through one stage of operation, which can be performed on either machine. After the completion

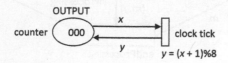

FIGURE 10.4
Final CPN model of the three-bit counter.

of processing, the part is unloaded from the system and a fresh part of the same type is loaded into the system. We define two sets as follows:

$$M = \{M1, M2\}$$

$$J = \{J1, J2, J3\}$$

M is the color set of machines, in which M1 is the color of the first machine and M2 the color of the second machine. J is the color set of parts, in which J1 is the color for parts of the first type, J2 the color of parts of the second type, and J3 the color of parts of the third type.

Figure 10.5 shows the CPN model of the manufacturing system. It has three places: Machine, Part, and InProcessing. As indicated in the model, the color set of Machine is M, the color set of Part is J, and the color set of InProcessing is M × J, a Cartesian product of M and J. In the initial marking, Machine has a token of the color M1 and a token of the color M2, meaning the two machines are available. Part has a token of the color J1, a token of the color of J2 and a token of the color J3, meaning a part of each type is ready for processing. There is no token in InProcess, meaning there is no part in processing in the initial state. The initial marking is formally denoted as

$$M_0(\text{Machine}) = 1`M1 + 1`M2 = M1 + M2$$

$$M_0(\text{Part}) = 1`J1 + 1`J2 + 1`J3 = J1 + J2 + J3$$

$$M_0(\text{InProcessing}) = \emptyset$$

Here, each of the three marking elements is expressed as a *multi-set*, which will be discussed more formally in the next section.

There are two transitions: beginProcessing and endProcessing. There are six arcs between transitions and places. A variable m is

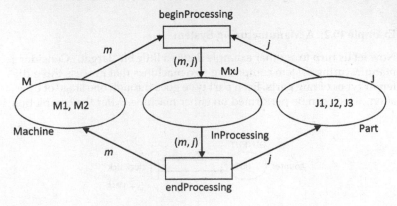

FIGURE 10.5
CPN model of a manufacturing system.

associated with the two arcs to/from Machine, a variable j is associated with the two arcs to/from Part, and a two-dimensional vector (m, j) is associated with the two arcs to/from InProcessing. Thus m binds to the color of tokens in Machine, j binds to the color of tokens in part, and (m, j) binds to the color of tokens in InProcessing.

In the initial marking, because there is no token in InProcessing, thus endProcessing is not enabled. beginProcessing, on the other hand, is enabled and can fire with respect to any of the following bindings:

$$<m = M1, j = J1>$$

$$<m = M1, j = J2>$$

$$<m = M1, j = J3>$$

$$<m = M2, j = J1>$$

$$<m = M2, j = J2>$$

$$<m = M2, j = J3>$$

Figure 10.6 shows the model after firing beginProcessing with respect to the first binding. In the new marking, only the token of the color M2 is left in Machine because the machine M1 is not available. For the same reason, the token J1 initially in part disappeared from the place. However, we get a new token (M1, J1) in the place InProcessing, indicating in this state, the first machine is processing a part of the first type.

In the new marking, both transitions are enabled. The transition beginProcessing can fire in any of the following two bindings:

$$<m = M2, j = J2>$$

$$<m = M2, j = J3>$$

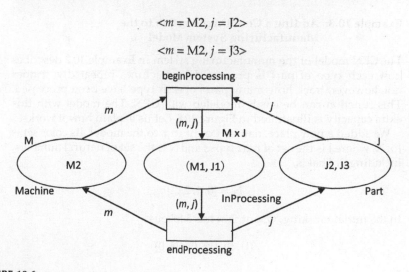

FIGURE 10.6

CPN model of the manufacturing system after firing beginProcessing with respect to the binding $<m = M1, j = J1>$.

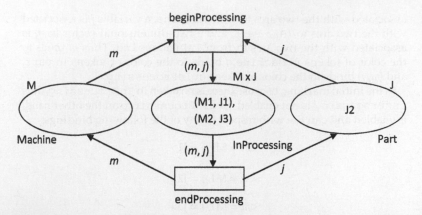

FIGURE 10.7
CPN model of the manufacturing system after firing `beginProcessing` with respect to the binding <m = M1, j = J1> and then firing `beginProcessing` with respect to the binding <m = M2, j = J3>.

and `endProcess` can fire in the following only binding:

$$<(m, j) = (M1, J1)>$$

Therefore, the next marking could be resulted in from three different transition firings. If `beginProcessing` in the second binding is selected to fire, then the resulting new marking is shown in Figure 10.7.

Example 10.3: Adding a Counting Function to the Manufacturing System Model

The CPN model of the manufacturing system in Example 10.2 describes how each type of part is processed by machines repeatedly. It does not, however, track how many parts of each type have been processed. This function can be easily modeled with CPN. The model with this extra capacity is illustrated in Figure 10.8. Let us explain how it works.

We added a new place, named `Counting`, to the model. Its color set is $J \times N$, where J is the set of part types and N is the set of natural numbers, including 0. That is,

$$N = \{0, 1, 2, \ldots\}$$

In the initial marking, `Counting` has 3 tokens:

$$(J1, 0), (J2, 0), (J3, 0)$$

That is,

$$M_0(\text{Counting}) = (J1, 0) + (J2, 0) + (J3, 0)$$

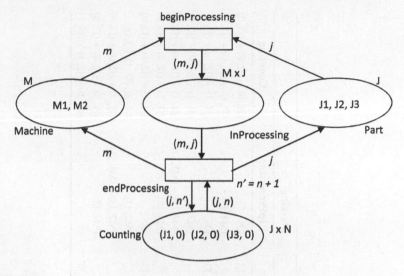

FIGURE 10.8
CPN model of the manufacturing system that counts how many parts of each type have been processed.

Each of the 3 tokens has two elements: the first element represents the part type, and the second element represents the number of parts of that type being processed. To the transition endProcessing, Counting is both an input place and an output place. In the initial marking, the input variable (j, n) binds to $(J1, 0)$, $(J2, 0)$, or $(J3, 0)$. Notice that the j in (j, n) must match the j in (m, j), therefore, when endProcessing fires, only the token in Counting that matches the token in InProcessing in terms of part type will be removed from Counting. The firing of endProcessing will also add a token back to Counting. Compared with the token removed from Counting, the added token has the same first element (part type) but an incremented second element (number of parts processed), which is enforced by the logic expression $n' = n + 1$ associated with the transition.

Table 10.1 lists a 10-step firing sequence and the change of markings after each step of transition firing. In the first step, we fire beginProcessing in the binding $<m = M1, j = J1>$. It results in a new marking, say M_1, as follows:

$$M_1(\text{Machine}) = M2$$

$$M_1(\text{InProcessing}) = (M1, J1)$$

$$M_1(\text{Part}) = J2 + J3$$

$$M_1(\text{Counting}) = (J1, 0) + (J2, 0) + (J3, 0)$$

TABLE 10.1

Execution of the CPN in Figure 10.8

	Firing Transition	Transition Binding	Machine	Part	Marking		Counting
					InProcessing		
0	–	–	M1 + M2	J1 + J2 + J3	∅		(J1, 0) + (J2, 0) + (J3, 0)
1	beginProcessing	⟨m = M1, j = J1⟩	M2	J2 + J3	(M1, J1)		(J1, 0) + (J2, 0) + (J3, 0)
2	endProcessing	⟨m = M1, j = J1, (j, n) = (J1, 0)⟩	M1 + M2	J1 + J2 + J3	∅		(J1, 1) + (J2, 0) + (J3, 0)
3	beginProcessing	⟨m = M2, j = J1⟩	M1	J2 + J3	(M2, J1)		(J1, 1) + (J2, 0) + (J3, 0)
4	beginProcessing	⟨m = M1, j = J2⟩	∅	J3	(M2, J1) + (M1, J2)		(J1, 1) + (J2, 0) + (J3, 0)
5	endProcessing	⟨m = M2, j = J1, (j, n) = (J1, 1)⟩	M2	J1 + J3	(M1, J2)		(J1, 2) + (J2, 0) + (J3, 0)
6	endProcessing	⟨m = M1, j = J2, (j, n) = (J2, 0)⟩	M1 + M2	J1 + J2 + J3	∅		(J1, 2) + (J2, 1) + (J3, 0)
7	beginProcessing	⟨m = M2, j = J3⟩	M1	J1 + J2	(M2, J3)		(J1, 2), (J2, 1), (J3, 0)
8	beginProcessing	⟨m = M1, j = J1⟩	∅	J2	(M2, J3) + (M1, J1)		(J1, 2) + (J2, 0) + (J3, 0)
9	endProcessing	⟨m = M1, j = J1, (j, n) = (J1, 2)⟩	M1	J1 + J2	(M2, J3)		(J1, 3) + (J2, 0) + (J3, 0)
10	endProcessing	⟨m = M2, j = J3, (j, n) = (J3, 0)⟩	M1 + M2	J1 + J2 + J3	∅		(J1, 3) + (J2, 1) + (J3, 1)

From this new state, both beginProcessing and endProcessing are firable. If we fire endProcessing as the second step, we will get a new marking, say M_2, as follows:

$$M_2(\text{Machine}) = M1 + M2$$

$$M_2(\text{InProcessing}) = 0$$

$$M_2(\text{Part}) = J1 + J2 + J3$$

$$M_2(\text{Counting}) = (J1, 1) + (J2, 0) + (J3, 0)$$

Notice that now the place Counting has a token that indicates one part of type J1 is processed. As shown in Table 10.1, if we continue to fire the two transitions according to the sequence listed in the second column and with respect to the bindings listed in the third column, the tokens in Counting will change as listed in the last column, which always records in a timely manner how many parts of each type have been processed.

10.2 Colored Petri Nets

After reading through the three examples given in the previous section, one should have some basic idea about CPNs and even why we need CPNs. Like regular P/T-nets, in addition to graphical representations, CPNs have a mathematical definition and formal semantics as well. Before we formally define CPNs, we introduce some concepts that will be used in the CPN definition.

10.2.1 Multi-Set

A *multi-set* m over a nonempty set S is defined as a function: $S \to N$, and represented as a sum:

$$\sum_{s \in S} m(s)\text{`}s$$

where $m(s)$ is the number of the element s in the multi-set. For example, for $S = \{a, b, c\}$, the following sums are all multi-sets over S:

$$2\text{`}a + 5\text{`}b + 1\text{`}c$$

$$1\text{`}a + 2\text{`}b$$

$$3\text{`}a$$

All the terms $1\grave{}e$, where e is an element of a multi-set, can be written as e. For example, the first multi-set listed above can be equally written as

$$2\grave{}a + 5\grave{}b + c$$

The set of all multi-sets over S is denoted by S_{MS}. Moreover, we use $\varnothing$ to represent an empty multi-set.

Assume m_1 and m_2 are two multi-sets over S. Operations of addition, subtraction, scalar multiplication, and comparisons are defined as follows:

$$m_1 + m_2 = \sum_{s \in S} \left(m_1(s) + m_2(s) \right)\grave{}s$$

$$m_1 - m_2 = \sum_{s \in S} \left(m_1(s) - m_2(s) \right)\grave{}s$$

$$n * m = \sum_{s \in S} \left(n * m(s) \right)\grave{}s$$

$$m_1 \neq m_2 \ \ iff \ \exists s \in S, m_1(s) \neq m_2(s)$$

$$m_1 \leq m_2 \ \ iff \ \forall s \in S, m_1(s) \leq m_2(s)$$

For example, if $m_1 = 2\grave{}a + 5\grave{}b + 1\grave{}c$ and $m_2 = 1\grave{}a + 2\grave{}b$, then we have

$$m_1 + m_2 = 3\grave{}a + 7\grave{}b + 1\grave{}c$$

$$m_1 - m_2 = 1\grave{}a + 3\grave{}b + 1\grave{}c$$

$$3 * m_2 = 3\grave{}a + 6\grave{}b$$

$$m_1 > m_2$$

10.2.2 Variable Set of an Expression

The set of variables in an expression *expr* is denoted by Var(*expr*). For example, the endProcessing transition is associated with an expression

$$expr: n' = n + 1$$

For this expression, we have Var(*expr*) = {n', n}.

10.2.3 Evaluation of an Expression

The value obtained by evaluating an expression *expr* in a binding b is denoted by *expr*<b>. Var(*expr*) is required to be a subset of the variables of b, and the evaluation is performed by substituting for each variable $v \in \text{Var}(expr)$ the value $b(v) \in \text{Type}(v)$ determined by the binding, where $\text{Type}(v)$ is a set of *types*, or colors, of variables v. An expression without variables is said to be a *closed* expression. It can be evaluated in all bindings, and all evaluations give the same result.

Mathematically, a CPN is a tuple $CPN = (P, T, \Sigma, C, G, A, E, I)$, where

1. $P = \{p_1, p_2, \ldots, p_m\}$ is a finite set of places.

2. $T = \{t_1, t_2, \ldots, t_n\}$ is a finite set of transitions such that $P \cup T \neq \emptyset$ and $P \cap T = \emptyset$.

3. Σ is a finite set of nonempty types, also called *color set*. A color set is essentially of a set of *data types*.

4. C is a color function. It is defined from P to Σ. $C(p)$ is the color set of p, such that

   a. $C(p) \subseteq \Sigma$

   b. $\underset{p \in P}{\cup} C(p) = \Sigma$

5. G is a guard function. It is defined from T into expressions such that $\forall t \in T$:

   a. $G(t)$ is a Boolean expression.

   b. $\text{Type}(\text{Var}(G(t))) \subseteq \Sigma$, that is, all variables in the expression take values from the color set.

6. A is a finite set of arcs. $A \subseteq P \times T \cup T \times P$.

7. E is an arc expression function. It is defined from A to expressions, such that $\forall a \in A$:

   a. $\text{Type}(E(a)) = C(p)_{MS}$, where p is the adjacent place of a.

   b. $\text{Type}(\text{Var}(E(a))) \subseteq \Sigma$, that is, all variables in the expression take values from the color set.

8. I is an initialization function that generates the initial marking. It is defined from P into closed expressions such that:

$$\forall p \in P: \text{Type}(I(p)) = C(p)_{MS}$$

A marking of a CPN is a function M defined on P such that for $p \in P, C(p) \rightarrow N$. Thus a marking M is an $n \times 1$ vector with components $M(p)$, where $M(p)$ represents the marking of place p and is represented by the formal sum of colors:

$$M(p) = \sum_{i=1}^{u} n_i \grave{} c_i$$

where n_i is the number of tokens of color c_i in the place p, that is,

$$M(p)(c_i) = n_i$$

u is the size of the color set of p, that is, $u = |C(p)|$. For example, let $\{c_1, c_2, c_3\}$ be the color set of a place p. In a marking M, p has 2 tokens of the color c_1, 1 token of the color c_2, and 0 tokens of the color c_3. Then we have

$$M(p) = 2\text{`}c_1 + 1\text{`}c_2 + 0\text{`}c_3$$

$$= 2\text{`}c_1 + 1\text{`}c_2$$

Define the set of arcs of a transition t as

$$A(t) = \{a \,|\, a \in P \times \{t\} \cup \{t\} \times P\}$$

Also define the set of variables of a transition t as

$$\text{Var}(t) = \{v \,|\, v \in \text{Var}(G(t)) \cup \exists a \in A(t) : v \in \text{Var}(E(a))\}$$

Then a *binding* of a transition t is a function b defined on $\text{Var}(t)$, such that

1. $\forall v \in \text{Var}(t) : b(v) \in \text{Type}(v)$
2. $G(t)<b>$ is true

The set of all bindings of t is denoted by $B(t)$. As shown in the previous section, bindings are often written in the form $<v_1 = c_1, v_2 = c_2, ..., v_n = c_n>$, where $\text{Var}(t) = \{v_1, v_2, ..., v_n\}$.

Denote by $E(p, t)<b>$ the evaluation of the expression defined on the arc (p, t) in the binding b. A transition t is said to be enabled with respect to a binding b in a marking M if and only if

$$M(p) \geq E(p, t)<b>, \forall p \in P \tag{10.1}$$

When a transition is enabled it may fire, changing the marking M_i to a different marking M_j, which is determined by

$$M_j(p) = M_i(p) - E(p, t)<b> + E(t, p)<b>, \forall p \in P \tag{10.2}$$

The definitions and notations about CPNs seem overwhelming. The following example should help you to gain a better understanding.

Example 10.4: CPN Definition

In this example, we use the notations that were introduced above to specify the CPN in Figure 10.8. The sets of places, transitions, and arc are very straightforward:

$P = \{$Machine, Part, InProcessing, Counting$\}$
$T = \{$beginProcessing, endProcessing$\}$
$A = \{($Machine, beginProcessing$)$, $($Part, beginProcessing$)$,
    $($beginProcessing,   InProcessing$)$, $($InProcessing,
      endProcessing$)$,
    $($Counting, endProcessing$)$, $($endProcessing, Counting$)\}$

The color set Σ of the CPN is the union of each place's color set, that is,

$$\Sigma = C(\text{Machine}) \cup C(\text{Part}) \cup C(\text{InProcessing}) \cup C(\text{Counting}),$$

where
$C(\text{Machine}) = \{$M1, M2$\}$,
$C(\text{Part}) = \{$J1, J2, J3$\}$,
$C(\text{InProcessing}) = \{($M1, J1$)$, $($M1, J2$)$, $($M1, J3$)$,
$($M2, J1$)$, $($M2, J2$)$, $($M2, J3$)\}$,
$C(\text{Counting}) = \{($J1, $n)$, $($J2, $n)$, $($J3, $n)\}$, $n \in \{0, 1, 2, \ldots\}$.

Notice that $($J1, $n)$ is not a single color; it is an aggregation of all two dimension colors that has J1 as the first element and any natural number as the second element. So are $($J2, $n)$ and $($J3, $n)$.

There is only one transition guard in this model, which is

$$G(\text{endProcessing}): n' = n + 1$$

This guard always evaluates to true as n is always valid regardless of what the type of parts is. It simply specifies the value of n'.

The arc expressions of this model are simply variables on arcs. They are

$E($Machine, beginProcessing$)$: m
$E($Part, beginProcessing$)$: j
$E($beginProcessing, InProcessing$)$: (m, j)
$E($InProcessing, endProcessing$)$: (m, j)
$E($Counting, endProcessing$)$: (j, n)
$E($endProcessing, Counting$)$: (j, n')

The initial marking of the model is as follows:

$$M_0(\text{Machine}) = \text{M1} + \text{M2}$$

$$M_0(\text{Part}) = \text{J1} + \text{J2} + \text{J3}$$

$$M_0(\text{InProcessing}) = \varnothing$$

$$M_0(\text{Counting}) = (\text{J1}, 0) + (\text{J2}, 0) + (\text{J3}, 0)$$

The binding set of each of the two transitions is as follows:

$B($beginProcessing$)$

$$= \{<m = \text{M1}, j = \text{J1}>, <m = \text{M1}, j = \text{J2}>, <m = \text{M1}, j = \text{J3}>,$$

$$<m = \text{M2}, j = \text{J1}>, <m = \text{M2}, j = \text{J2}>, <m = \text{M2}, j = \text{J3}>\},$$

B(endProcessing)

$$= \{<(m, j) = (M1, J1), (j, n) = (J1, n)>, <(m, j) = (M1, J2), (j, n) = (J2, n)>,$$
$$<(m, j) = (M1, J3), (j, n) = (J3, n)>, <(m, j) = (M2, J1), (j, n) = (J1, n)>,$$
$$<(m, j) = (M2, J2), (j, n) = (J2, n)>, <(m, j) = (M2, J3), (j, n) = (J3, n)>$$
$$| n \in \{0, 1, 2, \ldots\}\}$$

Now let us see how to formally check if a transition is enabled and calculate the new marking after a transition fires. Consider the transition beginProcessing and binding $<m = M1, j = J1>$. For the two input places Machine and Part, we have

$$M_0(\text{Machine}) = M1 + M2$$

$$E(\text{Machine, beginProcessing})<m = M1, j = J1> = M1$$

$$M_0(\text{Part}) = J1 + J2 + J3$$

$$E(\text{Part, beginProcessing})<m = M1, j = J1> = J1$$

Because

$$M_0(\text{Machine}) > E(\text{Machine, beginProcessing})<m = M1, j = J1>$$

$$M_0(\text{Part}) > E(\text{Part, beginProcessing})<m = M1, j = J1>$$

we know beginProcessing is enabled in the initial marking in the binding $<m = M1, j = J1>$.

After firing beginProcessing, the new marking, say M_1, should be calculated using Equation 10.2. The only output place of beginProcessing is InProcessing. Because

$$E(\text{beginProcessing, InProcessing})<m = M1, j = J1> = (M1, J1),$$

therefore,

$$M_1(\text{Machine}) = M1 + M2 - M1 = M2$$

$$M_1(\text{Part}) = J1 + J2 + J3 - J1 = J1 + J2$$

$$M_1(\text{InProcessing}) = \emptyset + (M1, J1) = (M1, J1)$$

This formal calculation result is consistent with our analysis in Example 10.3.

10.3 Analysis of Colored Petri Nets

CPNs have all the behavioral and structural properties that regular P/T-nets possess. CPN models can be analyzed based on state spaces or reachability graphs. The basic idea behind reachability graphs is to construct a directed graph that has a node for each reachable system state (marking) and an arc for each possible state change. To develop such a reachability graph, we first create a table like Table 10.1 to list all state changes and firing transitions and color bindings, and then draw the graph. Obviously, such a graph may become very large, even for small CPNs. However, it can be constructed and analyzed automatically with computer tools.

Given the inherent complexity of CPNs, simulation is always a good choice for the analysis of CPN models. The most outstanding CPN tool, called CPN Tools, was originally developed by the CPN Group at Aarhus University, Denmark, led by Prof. Kurt Jensen. In 2010, the tool was transferred to the AIS Group at Eindhoven University of Technology, Netherlands. CPN Tools is a tool for editing, simulating, and analyzing high-level Petri nets. It supports basic Petri nets plus timed Petri nets and colored Petri nets. It has a simulator and a state space analysis tool is included. The tool can be downloaded from http://cpntools.org/. It supports both Windows and Linux/Mac OS.

We can also perform invariant analysis on CPNs. This method is very similar to the use of invariants in ordinary program verification. The user constructs a set of equations which is proved to be satisfied for all reachable system states. For example, we know a manufacturing machine is either idle or busy, and thus we should have an equation as follows:

$$M(\texttt{Machine}) + M(\texttt{InProcessing})\texttt{<M>} = M1 + M2$$

where M is an arbitrary reachable marking and $M(\texttt{InProcessing})\texttt{<M>}$ indicates the element of the type M in the expression of $M(\texttt{InProcessing})$. By the same token, we should also have

$$M(\texttt{Part}) + M(\texttt{InProcessing})\texttt{<J>} = J1 + J2 + J3$$

If the above properties are not satisfied, then something is wrong with the modeling.

Exercises

1. Consider the simple CPN in Figure 10.9. The initial marking is

$$M_0(PA) = 2\grave{}a + b + 3\grave{}c$$

$$M_0(PB) = 3\grave{}d + 6\grave{}e$$

$$M_0(PA) = \varnothing$$

The binding set of the transition is

$$B(t) = \{<x = a, y = 2\grave{}d>, <x = b, y = 3\grave{}e>, <x = a, y = 2\grave{}e>, <x = 2\grave{}c, y = e>\}$$

Calculate the new markings after the transition fires in the bindings $<x = a, y = 2\grave{}d>$, $<x = a, y = 2\grave{}e>$, $<x = b, y = 3\grave{}e>$, and $<x = 2\grave{}c, y = d>$, sequentially.

2. Consider the simple CPN in Figure 10.10. The initial marking is

$$M_0(PA) = 6\grave{}a + 5\grave{}b + 4\grave{}c$$

$$M_0(PB) = \varnothing$$

$$M_0(PC) = \varnothing$$

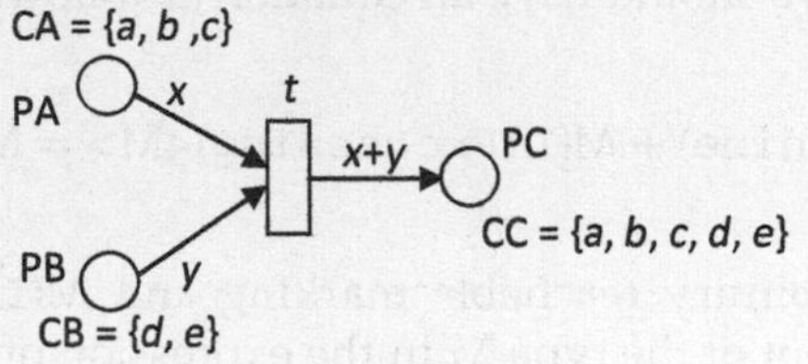

FIGURE 10.9
The CPN for Problem 1.

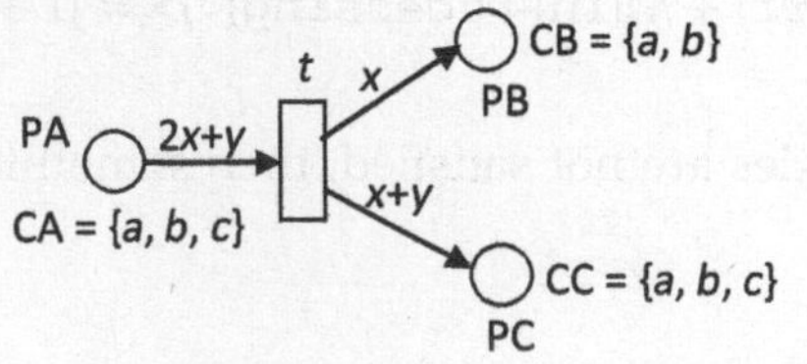

FIGURE 10.10
The CPN for Problem 2.

The binding set of the transition is

$$B(t) = \{<x = a, y = b>, <x = a, y = 2\hat{\ }c>,$$

$$<x = b, y = 2\hat{\ }c>\}$$

Calculate the new markings after the transition fires in the bindings $<x = a, y = b>$, $<x = a, y = 2\hat{\ }c>$, and $<x = b, y = 2\hat{\ }c>$, sequentially.

3. CPNs allow compact modeling, but they are equally expressive as regular Petri nets, provided that the set of token values is finite. Please convert the CPN model in Figure 10.5 to its equivalent regular P/T-net model.

4. Assume in a computer system there are N periodic processes that share a single processor. These processes are scheduled to run on the processor by a round-robin scheduler, as illustrated in Figure 10.11. That is, the system maintains a queue of the processes. Each time only the process at the head of the queue gets executed on the processor. When it is finished, the process will be put at the end of the queue. Meanwhile, the new head process is selected to run. Model the operation of these periodic processes with a CPN and give the initial marking of the model.

   Hint: Use a place for all processes and a separate place to indicate the process that has the privilege to run.

5. Consider the classic dining philosophers problem described in Problem 8, of the Chapter 7 Exercises, but we take off the assumption that one philosopher is an altruist. Thus when a philosopher gets his left fork, he will simply wait for his neighbor to return the fork on his right side to the table.

   a. Model the behavior of the philosophers with a regular P/T-net.
   b. Model the behavior of the philosophers with a CPN.
   c. List the color set of each place and binding set of each transition in the CPN model.

6. A traffic light set has three lights: one green, one yellow, and one red. During normal operation, the light color repeatedly changes from green to yellow, from yellow to red, and from red back to green.

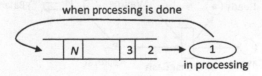

FIGURE 10.11
Periodic processes executed on a single processor scheduled with round-robin.

a. Model the color changes with a regular P/T-net.

b. Model the color changes with a CPN that has one place and three transitions.

c. Model the color changes with a CPN that has one place and one transition.

7. Figure 10.12 depicts the regular P/T-net model of a writing-reading system. The transitions write and send can keep firing and injecting as many tokens (mail) as you want to the place mail that is connected to send and receive. Therefore, this model assumes an unbounded buffer or mailbox for mail between the sender and receiver.

a. Consider the case of two writers and one reader. The two writers write and send mail to the same mailbox independently, and for each piece of mail the reader can identify who wrote it. The reader always receives two pieces of mail a time, one from each writer. If there is only one piece of mail in the mailbox, or all pieces of mail in the mailbox are from the same writer, the receiver won't take any of them. Please model the new system with a CPN. Notice that your CPN model should have the same structure as the one in Figure 10.12, but with color sets and arc expressions.

b. On top of (a), add a restriction to the mailbox capacity, such that the mailbox can hold up to four pieces of mail from each writer. You can only add one more place.

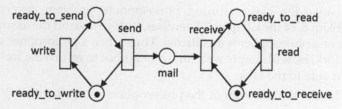

FIGURE 10.12
A writer and reader system.

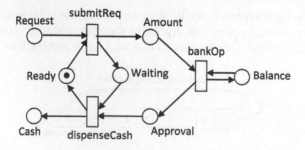

FIGURE 10.13
ATM operation.

8. The net in Figure 10.13 attempts to model a simplified operation of an automated teller machine (ATM) and its interaction with the bank behind it. When the ATM is in the ready state, a client can ask for a certain amount of cash. The ATM communicates this amount to the bank and waits for approval. When the approval arrives, the money is given to the client. Meanwhile, the requested amount is deducted from the client's account.

   a. Complete the model with all necessary elements to make it a complete CPN model. Make sure you define a guard for the transition bandOp in addition to all other elements.

   b. With this topologic structure (transitions, places, and arcs), if the amount requested by a client exceeds his or her balance in the bank, approval will not be given, and it leads to a deadlock. Modify the model such that in this case a non-approval message would be sent from the bank to the ATM, leading to an error message from the ATM to the client.

Index